Synopsis of Animal Classification

Synopsis of
Animal Classification

R. B. CLARK and A. L. PANCHEN

Department of Zoology
University of Newcastle upon Tyne

CHAPMAN AND HALL
LONDON

First published 1971
by Chapman and Hall Ltd
11 New Fetter Lane, London EC4P 4EE

First issued as a Science Paperback 1974

© 1971 R. B. Clark and A. L. Panchen

Photoset in Malta
by St Paul's Press Ltd
Printed in Great Britain by
Fletcher & Son Ltd, Norwich

ISBN 0 412 21250 1

Distributed in the U.S.A.
by Halsted Press, a Division
of John Wiley & Sons, Inc., New York

Contents

Preface   *page* vii

1   The Classification of Animals   1

2   The Animal Phyla   6

3   Classification of 'Non-Metazoan' Animals   8

4   Classification of Coelenterates   14

5   Classification of Acoelomates   22

6   Classification of Pseudocoelomates   30

7   Classification of Protostome Coelomates   36

8   Classification of Lophophorate Coelomates   77

9   Classification of Deuterostome Coelomates   80

Index   119

Preface

With the steady widening of the scope of zoology, the classical discipline of systematics receives scantier and scantier treatment in undergraduate courses, and the time when anyone claimed to be able to name almost any animal on sight is long since past. Nevertheless, it is still necessary for a zoologist to acquire as early in his career as possible some understanding of the classificatory system if only to make intelligible his studies in comparative physiology and similar disciplines.

It was as a solution to the practical problem of inculcating undergraduates with a minimal but adequate knowledge of systematics that this book was evolved in the Department of Zoology at Newcastle upon Tyne. It is impossible within the limited time of an undergraduate programme to treat modern zoology comprehensively and at the same time retain the systematic treatment of the animal kingdom that once formed the backbone of a zoology course. Greater and greater compression of such a systematic course gives the worst of all worlds: classical zoology is so inadequately treated that the course is almost worthless yet it still takes a disproportionate share of the undergraduate programme.

In seeking a solution to this dilemma we were forced to re-examine the function the systematic part of the course was supposed to serve. It proved to be twofold: firstly, to enable the student to name and place a large number of animals and, secondly, to become familiar with the principles of comparative functional morphology. In practice it was less successful in the latter than the former function. The same functional and anatomical principles were often discussed several times over in the context of different groups of animals, and not only was this time-wasting it also tended to obscure the underlying principles we were most anxious to illustrate.

The outcome of this re-examination was a decision to divide the course according to its two functions. The learning of the names of a number of animals in a taxonomic framework could then be regarded as acquiring a basic zoological vocabulary. Having done this, the student is in a position to add to his vocabulary as the need arises and to learn more about the animals he can now name.

The basic task can best be done in the laboratory where the animals can be seen and handled, and since the immediate aim is chiefly to acquire a vocabulary of names, the operation is much like that of learning aircraft recognition. From the outset, however, the student must be given some clues as to what are significant features by which animal groups can be recognized, and for that purpose brief diagnostic notes were provided on cards beside the specimens. Later, students were given printed sheets which could be taken away and finally these were collected together in book form. It is from the latter that this published version has developed.

In deciding the form and content of this synopsis we have tried to bear in mind the needs of its potential users.

Diagnoses of taxa are given down to the rank of order or very occasionally, sub-order. At this level all the major morphological types of animal are distinguished and beyond it one is concerned with detail which is properly the concern of the specialist.

As far as conveniently possible, diagnoses have been framed in non-technical terms. Where this has been impossible, the terms have been defined. In general the language is that of the student zoologist who lacks any detailed knowledge of animal morphology and the terminology associated with it. Sometimes it has been simpler to provide a drawing than to give a lengthy description, but while there might have been some advantage to providing a larger number of illustrations, to have done that would have posed serious problems of selection while extending the scope of the book beyond the limits we have set ourselves.

Finally, we have had to face severe problems in selecting acceptable and useful systems of classification. For some groups we have had to compromise between the most recent acceptable classification and the one in wide use in textbooks. It is too much to hope that the classification we have used will please everyone but it has seemed most important to allow for easy cross reference between this synopsis and standard textbooks.

To meet the needs of courses on vertebrate anatomy and evolution, the classification of the group includes that of extinct forms, but otherwise, as a general rule, only living animals are included.

We are grateful to our colleagues in the Department of Zoology in this University for their assistance and advice in the preparation of this synopsis.

1 The Classification of Animals

Animals are classified in a hierarchical system. The animal KINGDOM is divided into three SUB-KINGDOMS, Parazoa, Mesozoa and Metazoa. (The Protozoa may be regarded as a fourth sub-kingdom or included in a separate kingdom, the Protista; they are not treated here.) The largest of these three sub-kingdoms, the Metazoa, is divided into PHYLA, while the other two sub-kingdoms contain only a single phylum each. Phyla are subdivided into CLASSES, and so on. The major categories (taxa) are as follows:

> KINGDOM
> PHYLUM
> CLASS
> ORDER
> FAMILY
> GENUS
> SPECIES

Definitions of phyla are based on major structural differences between different kinds of animals and members of each phylum are built on a distinct structural plan. Consequently the number of animals in each phylum varies considerably. Particularly in the larger phyla it is convenient to introduce additional categories, e.g. SUB-PHYLUM, SUB-CLASS, SUPER-FAMILY. Usage of these secondary taxa may vary slightly with different writers. Thus, the lungfish may be designated order Dipnoi of the sub-class Sarcopterygii in the class Osteichthyes (bony fish), or sub-class Dipnoi, class Osteichthyes, or class Dipnoi, super-class Osteichthyes, all within the sub-Phylum Craniata (Vertebrata).

Ideally, the classificatory system should reflect the evolutionary relationships between animals, so that all the members of one taxon (e.g. a family) are more closely related to one-another than they are to members of a neighbouring taxon of equivalent rank (e.g. to members of another family in the same order). In practice we do not always know enough about the animals to be certain of their relationship to other animals and views may change from time to time. For immediate practical purposes it is essential to classify and name animals, even if this is on a provisional basis and known to be unsatisfactory. Very occasionally an animal or group of animals is so unusual that no conclusions can be made about its relationship to others. It is then designated *incertae sedis* meaning that its position in the classificatory system is uncertain. Because of the essentially provisional nature of much classification, the entire system is constantly reviewed and adjusted to take account of new information and new views. Fortunately, the major groups of animals are well known and most of the revision is at a detailed level so that it affects only the specialist.

The Principles of Taxonomic Procedure

Taxonomy is concerned with two things:
1. giving each species of animal a name;
2. arranging species in appropriate genera, families, orders and so on, so that the relationship between different animals is, so far as possible, indicated in the classificatory system.

If a hitherto unknown animal is discovered, it must be
1. described in sufficient detail to enable other zoologists to recognize it when other specimens are found;
2. given a name (the specimen or specimens from which the new species was described then become 'type specimens' of the species and are referred to in future cases of doubt);
3. assigned to a genus, family, order, etc., to indicate its relationship to other animals and, if necessary, new families, orders, etc., have to be created, named and defined, to accommodate it.

An elaborate system of regulations governing taxonomic procedure has been evolved to ensure consistent treatment by all taxonomists. The regulations are given in the International Rules of Zoological Nomenclature. What follows is an explanation of the more immediate and important applications of the Rules.

Animal Names

All animals are referred to by two names, those of the genus and species (or generic and specific names), comparable to surname and first name. Thus the frog common in Great Britain is *Rana temporaria*. The generic name *Rana* is always given a capital initial and placed before the specific name *temporaria* which has a small initial. If the full name has already been used in a paper and there is no chance of ambiguity, the name may be abbreviated to *R. temporaria*. Again, if there is no ambiguity, the frog may be referred to simply by its generic name *Rana*, although this does not distinguish between *R. temporaria*, *R. fusca*, *R. esculenta* and a dozen other closely related frogs in the same genus. It is conventional to print the scientific names of animals in italics and underline them in manuscript.

All names are latin or latinized and the species name must be treated grammatically as one of the following:

1. an adjective in the nominative singular agreeing with the gender of the generic name, e.g. *Rana esculenta*, *Echinus esculentus*;

2. a noun in the nominative singular in apposition to the generic name, e.g. *Panthera leo*;

3. a noun in the genitive case, e.g. *Anaspides tasmaniae*; this is often used for parasites in which the specific name is derived from the name of the host, e.g. *Athelges paguri*, an isopod parasitic on the hermit crab *Pagurus bernhardus*.

The species name should strictly be followed by the name of the author who gave it that name and by the date of publication of that name, but this practice is not invariably followed in the general zoological literature. If a species has subsequently been transferred to a different genus from the one to which the author originally assigned it, the author's name is shown in brackets, but not if the species is still in its original genus. Thus, *Spirontocaris occultus* Lebour is now *Eualus occultus* (Lebour).

Occasionally animals are given three names, the third being that of the sub-species. Clearly defined sub-species are geographical races, varieties, etc. and are recognized in only a few groups. Usage varies a good deal and the classificatory system is not well adapted to deal with categories below the level of species.

Taxonomic Revision

Theoretically, once an animal has been described, named and classified, that is the end of the matter. In practice, mistakes are sometimes made, or the discovery of a new species may render the existing system of classification obsolete, and a revision of the taxonomy becomes necessary.

Species name

Once a valid species name has been published, it cannot be changed even by the original author, but the name may be invalid for several reasons, the commonest of which are:

1. Unknown to the author, the species had already been discovered, adequately described and named by someone else. The earlier name normally takes precedence. An exception to this priority rule is made if the first name, although valid, has not been used for at least 60 years and the animal is already well known by another name.

2. The species occurs in several different forms (e.g. it becomes modified for breeding or passes through a variety of developmental stages) and each form was described as a different species before the relationship between them was appreciated. Again, the earliest valid name normally takes precedence.

3. The original description is so inadequate that it is impossible to know whether other specimens found subsequently should be referred to the species or not. This can be due either to the poor condition of the original specimen(s) (which is common in fossils) or to ambiguities in the original description.

Thus there may exist in the scientific literature several different names for the same animal. Only the earliest of these is valid, the others are synonyms.

By international agreement, this classificatory system is held to have begun in 1785 with the publication of the 10th edition of Linnaeus' *Systema naturae*, and names published before January 1785 cannot claim priority.

Classification

With an agreed starting date and priority being given to the earliest valid name given to a species after that date, the system of nomenclature at the species level should be stable and is so to a very great extent. The same priority is given to the earliest generic names and sometimes to family names, but because these taxa indicate the evolutionary relationships between animals and our views about them change as more information becomes available, we cannot expect the same degree of stability of generic and family names as of specific names.

Thus, a species may be transferred from one genus to another, genera previously regarded as separate may be merged into a single genus, or a genus may be split into two or more new genera. Genera may similarly be redistributed between families, and families split or merged, or new ones created. Comparable adjustments go on at all taxonomic levels, but changes in genera have the most immediate practical consequences because the generic name forms part of the name of the animal. (Such a change is comparable to a woman changing her surname on marriage.) With the transfer of a species to a new genus the specific name is basically unchanged, except for grammatical adjustment, e.g. of gender if the genus name now has a feminine ending instead of masculine.

The law of priority which governs the naming of species and genera does not apply and is usually not observed for taxa above the rank of genus.

2 The Animal Phyla

There is no agreed grouping of metazoan phyla in the nomenclature although several systems have been proposed at various times in the past. With about twenty-seven phyla in the sub-kingdom, however, it is convenient to have some classification of them even though the grouping may indicate evolutionary relationships only vaguely and doubtfully. The terms and groups used here are purely descriptive of the grades of organization of the animals. They may reflect evolutionary relationships but there is no guarantee of this and the terms have no official standing in the classificatory system. They are used only for convenience.

Some authorities do not regard the Parazoa and Mesozoa as constituting separate sub-kingdoms, but include them in the Metazoa. In this case, the Mesozoa are regarded as secondarily reduced rather than primitively simple and are tentatively associated with the Platyhelminthes. The Porifera remain a separate phylum but it is still recognized that they are rather different from all other metazoans which are separated from them as 'Eumetazoa.'

Sub-Kingdom I Parazoa
Phylum 1 Porifera

Sub-Kingdom II Mesozoa
Phylum 1 Mesozoa

Sub-Kingdom III Metazoa

(a) Coelenterates
Phylum 1 Cnidaria
Phylum 2 Ctenophora

(b) Acoelomates
Phylum 1 Platyhelminthes
Phylum 2 Rhynchocoela

(*c*) Pseudocoelomates
Phylum 1 Rotifera
Phylum 2 Gastrotricha
Phylum 3 Kinorhyncha
Phylum 4 Nematoda
Phylum 5 Nematomorpha
>(These five phyla are sometimes regarded as classes of a single, large phylum Aschelminthes.)

Phylum 6 Acanthocephala
Phylum 7 Endoprocta
Phylum 8 Priapulida

(*d*) Protostome coelomates
Phylum 1 Sipuncula
Phylum 2 Echiura
Phylum 3 Mollusca
Phylum 4 Annelida
Phylum 5 Tardigrada
Phylum 6 Pentastomida
Phylum 7 Onychophora
Phylum 8 Arthropoda

(*e*) Lophophorate coelomates
Phylum 1 Phoronida
Phylum 2 Brachiopoda
Phylum 3 Ectoprocta

(*f*) Deuterostome coelomates
Phylum 1 Chaetognatha
Phylum 2 Pogonophora
Phylum 3 Hemichordata
Phylum 4 Echinodermata
Phylum 5 Chordata

3 Classification of 'Non-Metazoan' Animals

The division of the kingdom Animalia into sub-kingdoms presents problems and any solution is in some degree arbitrary.

There is a strong possibility that the Mesozoa are not primitively simple in construction but are metazoans (probably related to the Platyhelminthes) that have become secondarily reduced with the adoption of a parasitic habit.

The position of the Porifera is also unclear. At one extreme they may be included in the Metazoa, at the other, separated into two phyla in an independent sub-kingdom Parazoa.

If both the Mesozoa and Porifera are included in the Metazoa, the need for sub-kingdoms vanishes (though even so the difference in order of complexity between sponges and other animals is recognized by designating all phyla except the Porifera as Eumetazoa).

On the other hand, in both the Parazoa and Mesozoa, the organization of cells is loose and labile and, unlike those of metazoans, they never constitute tissues. This is often taken to be a difference of fundamental significance and it is tempting then to regard Parazoa, Mesozoa and Metazoa as having independent origins from the non-cellular Protista. Another view, now rather out of favour, is that the Parazoa and Mesozoa are stepping stones in the evolution of the Metazoa from the Protista. The line of reasoning that separates the Metazoa from the other two groups of multicellular animals is the justification for placing them in distinct sub-kingdoms.

Summary Classification on 'Non-Metazoan' Animals

Phylum Porifera
    Sub-Phylum Nuda
        Class 1 Hexactinellida

Sub-Phylum Gelatinosa
   Class 1 Calcarea
   Class 2 Demospongea
      Sub-class 1 Tetraactinellida
      Sub-class 2 Monaxonida
      Sub-class 3 Keratosa
Phylum Mesozoa
   Class 1 Orthonectida
   Class 2 Dicyemida

PHYLUM PORIFERA

Asymmetrical or radially symmetrical multicellular animals, without organs, mouth or nervous system. Body penetrated by pores, canals and chambers through which water flows; some of the internal cavities are lined with characteristic flagellated collar cells (choanocytes).

The following classification, based on adult characters, is in widest use. It has a number of unsatisfactory features and is being replaced by alternative systems of classification which take into account the cytology and embryology of the animals. The effect of these changes is indicated where appropriate.

SUB-PHYLUM NUDA

Sponges without a true epidermis, external surface composed of a trabecular net derived from amoebocytes. No gelatinous layer.

Sometimes raised to the rank of a separate phylum, but equally, although the distinction between hexactinellids and other sponges is recognized, may be reduced to a class equivalent to the remaining classes.

Class Hexactinellida

Skeleton composed of six-rayed silica spicules, either separate or fused together to form a network. Choanocytes confined to finger-shaped chambers. No covering epithelium (epidermis). Body of the sponge generally radially symmetrical. The large spicules (megascleres) all six-rayed (Fig. 1) but may have equally or unequally developed rays. The small spicules (microscleres) more variable.

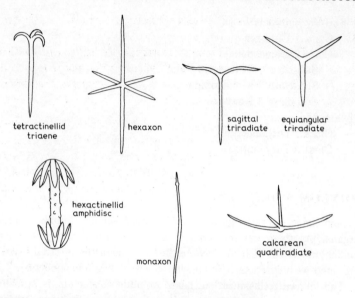

FIGURE 1   Sponge spicules (after Hartman and Hyman).

Order 1. **Hexasterophora.** Microscleres all basically six-rayed (hexastines) miniature versions of megascleres or with branched ends. Examples: *Euplectella, Farrea.*
Order 2. **Amphidiscophora.** Microscleres all amphidiscs (Fig. 1). Examples: *Monoraphis, Pheronema, Hyalonema.*

SUB-PHYLUM GELATINOSA

Sponges with a surface epithelial layer and a middle gelatinous layer containing amoebocytes, etc., corresponding to a mesogloea. May be raised to the status of a separate phylum or, at the other extreme, reduced to a class.

Class Calcarea
Sponges with a calcareous skeleton composed of 1-, 3-, or 4-rayed spicules.
   The following classification, based chiefly on the distribution of the choanocytes, is in common use.
Order 1. **Homocoela.** Simple sponges with the main central cavity

lined with choanocytes. Examples: *Clathrina, Dendya, Leucosolenia, Ascyssa.*

Order 2. Heterocoela. The main central cavity (spongocoel) without choanocytes which are confined to flagellated in outpocketings of the body wall. Examples: *Leucascus, Leucetta, Minchinella, Petrobiona, Scypha* (= *Sycon*), *Grantia, Leucilla.*

An alternative and more satisfactory classification by W. D. Hartman (*Syst. Zool.*, 1958, **7**, 97—110.) directs attention to cytological and embryological features and results in a redistribution of members of the Homocoela and Heterocoela in two sub-classes. On this basis it is understood that the homocoel and heterocoel distribution of choanocytes has been independently reached in the two sub-classes.

Sub-class Calcinea
Choanocytes with a basal nucleus, the flagellum arising independently of the nucleus. Parenchymula-like larva. Some or all of triradiate spicules, when present, equiangular and equiradiate (Fig. 1). Monaxons may be present or absent. Triradiates the first spicules to appear in the newly settled sponge.

Order 1. Clathrinida. Spongocoel lined with choanocytes throughout life. No true dermal membrane or cortex. Examples: *Clathrina, Dendya.*

Order 2. Leucettida. Spongocoel not lined with choanocytes which are confined to the flagellated chambers. Distinct dermal membrane or cortex. Examples: *Leucascus, Leucetta.*

Order 3. Pharetronida. Main skeleton of quadriradiate spicules (Fig. 1) joined by a calcareous cement, or comprising a calcareous network not composed of spicules. Sometimes regarded as a separate sub-class. Examples: *Minchinella, Petrobiona.*

Sub-class Calcaronea
Choanocytes with an apical nucleus, the flagellum arising directly from it. Ambiblastula larvae. Triradiate spicules not equiangular but with two of the rays almost in the line with one-another (sagittal) (Fig. 1). Monaxons usually present. Monaxons the first spicules to appear in the newly settled sponge.

Order 1. Leucosoleniidae. Spongocoel lined with choanocytes throughout life. No true dermal membrane or cortex. Examples: *Leucosolenia, Ascyssa.*

Order 2. Sycettida. Spongocoel not lined with choanocytes which are confined to the flagellated chambers. Continuous dermal membrane or cortex present (except in family Sycettidae). Examples: *Scypha* (= *Sycon*), *Grantia*, *Leucilla*.

Class Demospongea

Skeleton composed of siliceous spicules or horny fibres (spongin) or both. Spicules are not six-rayed and usually differentiated into large megascleres and small microscleres.

This classification, although currently widely used, is acknowledged to be artificial and unsatisfactory. An alternative, taking into account embryological features and reflecting evolutionary relationships more closely, has not yet come into use.

Sub-class Tetractinellida

Spicules 4-rayed (tetraxons). No spongin. Sometimes lack spicules altogether.

Order 1. Myxospongida. Structure simple, no spicules. Examples: *Halisarca*, *Oscarella*, *Hexidella*, *Bajulus*.

Order 2. Carnosa. Megascleres and microscleres not sharply differentiated. Examples: *Plakortis*, *Plakina*, *Chrondrilla*.

Order 3. Choristida. Tetraradiate spicules (Fig. 1) with one very long shaft and three small rays at the top (triaenes). Megascleres and microscleres sharply differentiated. Examples: *Craniella*, *Ancorina*, *Thenea*.

Sub-class Monaxonida

The commonest of all sponges. Megascleres are with 1 ray. With or without spongin fibres.

Order 1. Hadromerina. Megascleres mainly with a knob at one end, microscleres are in the form of spiny nodules or are star-shaped. Without spongin. Examples: *Suberites*, *Poterion*, *Cliona*, *Tethya*.

Order 2. Halichondrina. Megascleres generally of two or more kinds. With little spongin. Example: *Halichondria*.

Order 3. Poecilosclerina. Includes the majority of Demospongea. Megascleres often of two or more kinds and localized in their distribution, joined together by spongin fibres, and often projecting from the body. A variety of microsclere types. Examples: *Myxilla*, *Clodorhiza*, *Esperiopsis*.

Order 4. Haplosclerina. Megascleres all pointed at both ends and grow in both directions from a central point. With or without microscleres but generally with spongin fibres. Examples: *Spongilla* (in freshwater), *Haliclona.*

Sub-class Keratosa

Skeleton composed exclusively of spongin. No spicules. Examples: *Spongia, Hircinia, Aplysilla.*

PHYLUM MESOZOA

Small multicellular animals, all parasitic in marine organisms. The body composed of two layers of cells. Complex life history, passing through larval stages of quite different morphology from the adult. May not be correctly regarded as constituting a separate phylum, but should be included in Platyhelminthes.

Class Orthonectida

Body more or less annulated. An outer cell-layer enclosing a central mass composed entirely of germinal cells. The class includes only two genera in separate families. Examples: *Rhopalura, Permatosphaera.*

Class Dicyemida

Body not annulated. Living the greater part of the life cycle in the kidney of cephalopod molluscs.
Order 1. Dicyemida. Body entirely ciliated. Examples: *Dicyema, Pseudocyema, Dicyemennea* (only genera).
Order 2. Heterocyemida. Body ciliation completely lost in adult stage. The 10 ectodermal cells tend to fuse and form a syncytium. Examples: *Conocyema, Microcyema* (only genera).

4 Classification of Coelenterates

Metazoa with a primary radial symmetry. The body encloses a single cavity, the *coelenteron* or *gastrovascular cavity*, with a single opening to the exterior, the mouth. This cavity may sometimes be divided into canals, etc. Body wall composed of three layers: an outer *epidermis*, an inner *gastrodermis* lining the gastrovascular cavity, and the *mesogloea* between them. The mesogloea varies from a thin, noncellular cement to a thick gelatinous or fibrous layer containing cellular elements. Unlike the mesoderm which occupies a corresponding position between the inner and outer layers in all non-coelenterate metazoans, the mesogloea never produces organ structures and for this reason is not regarded as a true tissue-layer. Hence coelenterates are sometimes regarded as *diploblastic* animals. (i.e. with a body composed of only two tissue-layers), but the term is now outmoded although it is still useful in an imprecise way.

Summary Classification of Coelenterates

Phylum Cnidaria
    Class 1 Hydrozoa
    Class 2 Scyphozoa
    Class 3 Anthozoa
        Sub-class 1. Ceriantipatharia
        Sub-class 2. Octocorallia
        Sub-class 3. Zoantharia
Phylum Ctenophora
    Class 1 Tentaculata
    Class 2 Nuda

PHYLUM CNIDARIA

Characterized by the possession of unique stinging cells (nematocysts). Generally built on a radial plan with tentacles fringing the mouth. There are two basic morphological types, medusa (Fig. 2) and polyp (Fig. 3), both of which may occur in successive phases of the life history, or one or other may be dominant. There is a strong tendency to the development of colonial forms, sometimes with both morphological types being simultaneously represented.

Class Hydrozoa

Radially symmetrical (generally tetramerous or polymerous) cnidarians with polyp and medusa constituting important phases of the life cycle, although there is a tendency for the medusa to be retained on the polyp, reduced, and ultimately to disappear altogether. Polyp with a simple gastrovascular cavity not sub-divided or partitioned; medusa (Fig. 2) almost always craspedote (i.e. with a velum).

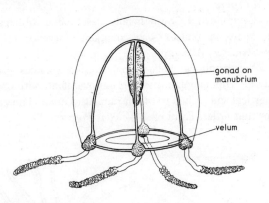

gonad on manubrium

velum

FIGURE 2   Hydrozoan medusa (*Sarsia*) (after Russell).

Classification of the Hydrozoa is bedevilled by the fact that although polyps and medusae are separately well known, in many cases it is not known which polyps and medusae belong to the same organism. Even when this is known, the two phases may still (improperly) have different generic and specific names. This confusion has extended to higher taxa and there have been what amounted to two separate classifications, one of polyps, the other of medusae, with very little correspondence between them. Textbooks on the whole have reflected the polyp classification. More recent systems of classification, as in Rothschild's (1965) *A Classification of Living Animals*: second edition (Longmans, London), which provides the basis for that given below, go some way towards removing the illogicalities. Correspondence between the current and older classifications is indicated whenever possible.

Orders Athecata, Thecata, Limnomedusae and Chondrophora are sometimes regarded as sub-orders of the order Hydroida.

Order 1. Athecata (= Anthomedusae). Medusa generally deep bell-shaped with gonads on the stomach or, very rarely, extending onto the sub-umbrella. With or without eyes, but without statocysts. Hydranths of polyps not in a hydrotheca, investing skeletal material confined to stalk (Fig. 3). Examples: *Hydractinea, Tubularia, Syncorine* (polyp), *Sarsia* (medusa), *Hydra*.

Order 2. Milleporina. Colonial forms with polyps living in pits in the surface of a massive calcareous skeleton secreted by the epidermis. Brief medusoid phase. Now commonly included in Athecata. Example: *Millepora*.

Order 3. Stylasterina. Superficially resembling Milleporina but lacking a medusoid phase. Also commonly included in Athecata. Examples: *Stylaster, Allopora*.

Order 4. Chondrophora. Pelagic, polymorphic colonies composed of polyps only (cf. Siphonophora) with a flat oval float; with or without a diagonal vertical sail. Allied to tubularian Athecata and now commonly included in that order. Examples: *Velella, Porpita*.

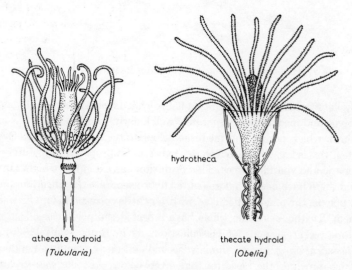

athecate hydroid
(Tubularia)

hydrotheca

thecate hydroid
(Obelia)

FIGURE 3   Hydrozoan polyps (after Hyman).

Order 5. Thecata (= Leptomedusae). Medusae with hemispherical or flattened umbrella, with gonads on the radial canals. Usually with statocysts. Polyps almost always have hydrotheca into which they can be withdrawn (Fig. 3). (In the very few cases where this is not so, loss of hydrotheca thought to be secondary.) Examples: *Obelia, Sertularia, Plumularia, Aequoria, Phialidium*.

Order 6. Limnomedusae. Medusae with hollow tentacles, gonads on stomach wall, with or without extensions along the radial canals, or on the radial canals only. If statocysts are present, they are internal and have an endodermal axis. Polyp often reduced, with or without tentacles, the endoderm of the tentacles in direct connection with that of gastrovascular cavity. Mostly freshwater forms that at one time were included in the Thecata, Athecata and Trachymedusae. Examples: *Gonionemus, Craspedacusta, Limnocnida*.

Note: the six foregoing orders may sometimes all be included in a large order Hydroida. In older classifications, this was done except that Milleporina and Stylasterina were separated as independent orders or as sub-orders of the order Hydrocorallina; the Chondrophora were often included in the Siphonophora.

Order 7. Trachymedusae. No polypoid phase. Medusa with a hemispherical or deep bell-shaped umbrella with the margin not divided into lobes but with a thickened, marginal nematocyst ring. Gonads usually on the radial canals. Marginal tentacles all solid, or both hollow and solid types present. Examples: *Liriope, Geryonia, Aglantha*.

Order 8. Narcomedusae. Medusae with solid tentacles which leave the umbrella some distance above its margin; a groove from the tentacle base to the margin is sometimes cleft so that the umbrella margin appears lobed. No radial canals and if a ring canal is present it is interrupted near the base of the tentacles. Examples: *Aegina, Solmissus, Cunina*.

Note: the two foregoing orders are sometimes regarded as sub-orders of the order Trachylina.

Order 9. Siphonophora. Very polymorphic, pelagic, colonial forms with both polypoid and medusoid forms existing simultaneously in the same colony. The first individual(s) in the colony modified for swimming or as a float. Examples: *Muggiaea, Physalia, Stephalia*.

Order 10. Actinulida. Very small, solitary hydrozoans modified for life as meiofauna. Retain into adult life the bipolar organization of the actinula and covered with cilia. Some with aboral adhesive organ. Examples: *Halammohydra, Otohydra* (only genera).

Class Scyphozoa

Medusoid phase free-swimming or attached by an aboral stalk, dominant. The polypoid phase, if present, small and temporary, developing directly into medusoids or budding them off by transverse fission. Gonads in the gastrodermis. Medusoid lacks a velum.

Order 1. Stauromedusae. Attached by an aboral stalk and developing directly from the polypoid phase. Examples: *Haliclystus*, *Lucernaria*, *Ceratolophus*.

Order 2. Cubomedusae. Free-swimming, cuboidal medusae with a tentacle or group of tentacles at each corner of the bell. The free edge of the bell is bent inwards to act as a velum. Examples: *Carybdea*, *Chironex*.

Order 3. Coronatae. Free-swimming medusae with the edge of the bell scalloped and separated from the rest of the bell by a furrow. Examples: *Nausithoë*, *Periphylla*, *Atolla*.

Order 4. Semaeostomeae. Corners of the mouth prolonged as four frilly lobes. No tentacles. Examples: *Aurelia*, *Chrysaora*, *Cyanea*, *Pelagia*.

Order 5. Rhizostomeae. The oral lobes fused together, obliterating the mouth, but numerous small mouths formed on the oral lobes. No tentacles. Examples: *Rhizostoma*, *Cassiopeia*, *Mastigias*, *Stomalophus*.

Class Anthozoa

Polypoid dominant, medusoid almost completely absent. The region around the mouth of the polyp is expanded to form an oral disc. Mouth opens into a sleeve-like stomodeum leading into the gastrovascular cavity which is partitioned by a series of radial mesenteries or septa. Gonads are gastrodermal and located on the septa. Mesogloea contains cells and is often well developed.

Sub-class Ceriantipatharia

Primitive anthozoans formerly included in the Zoantharia. Simple arrangement of septa and tentacles, musculature unspecialized and no differentiation of elements within the mesogloea. Septa arranged in bilaterally symmetrical couples (Fig. 4). Primary septa (generally 6) complete and if further septa are formed these are added sequentially at the ventral side of the polyp.

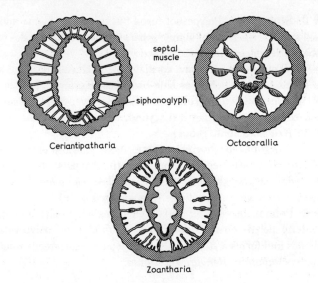

septal muscle

siphonoglyph

Ceriantipatharia

Octocorallia

Zoantharia

FIGURE 4   Arrangement of septa in the Anthozoa (after Hyman).

Order 1. Antipatharia (black corals). Colonial, with a thorny black or brown axial skeleton bearing the polyps laterally. 6, 10 or 12 (generally 10) complete septa arranged in couples. Generally 6 tentacles which are continuous with the interseptal spaces between the 6 primary septa. Example: *Antipathes*.

Order 2. Ceriantharia. Long, solitary anemones without a pedal disc. Two rings of tentacles. Single dorsal siphonoglyph. 6 primary septa in couples, indefinite number of couples of further septa added ventrally and alternate long and short. Examples: *Cerianthus*, *Arachnactis*.

Sub-class Octocorallia (= Alcyonaria)

Colonial anthozoans with 8 pinnate tentacles and 8 complete septa (Fig. 4). 1 ventral siphonoglyph. If a skeleton is formed it is internal and generally composed of spicules. The first four orders are commonly merged in the Alcyonacea.

Order 1. Alcyonacea (soft corals). Lower parts of polyps fuse to form soft mass from which oral ends protrude. Calcareous spicules laid down in mass as a skeleton. Examples: *Alcyonium*, *Gersemia*.

Order 2. Stolonifera. Polyps not fused but connected to one-another by basal stolons. Skeleton of simple separate spicules, or spicules forming tubes. Examples: *Tubipora, Clavularia*.

Order 3. Telestacea. Colony consists of a cluster of long axial polyps bearing lateral polyps as side-branches. Example: *Telesto*.

Order 4. Coenothecalia (blue coral). Massive, solid calcareous skeleton containing long, vertical, cylindrical cavities for the polyps. Example: *Heliopora* (only genus).

Order 5. Gorgonacea (horny corals). Axial skeleton of calcareous spicules, or horny secreted material, or both. Polyps short and borne on the sides of the skeletal axis. Examples: *Gorgonia* (sea fan), *Leptogorgia* (sea whip), *Corallium* (precious red coral).

Order 6. Pennatulacea (sea pens). One long axial polyp with many short lateral polyps on its sides. Lower part of axial polyp without lateral ones and forms a stalk. Skeleton of separate calcareous spicules. Examples: *Pennatula, Renilla, Veretillum*.

Sub-class Zoantharia

Solitary or colonial anthozoans. With septa in multiples of 6 as a rule (Fig. 4) and arranged in pairs. Tentacles not pinnate. If a skeleton is formed, it is external and not composed of spicules.

Order 1. Zoanthiniaria (= Zoanthidea). Small, solitary or colonial anthozoans without a skeleton or pedal disc. A single, ventral siphonoglyph. Septa in pairs, one complete, the other incomplete. Many epizoic on other invertebrates. Examples: *Zoanthus, Epizoanthus, Polythoa*.

Order 2. Corallimorpharia. Resembling true corals, though without a skeleton. Tentacles radially arranged. Example: *Corynactis*.

Order 3. Actiniaria (sea anemones). Solitary anthozoans without a skeleton, the aboral end generally forming an attachment disc. Usually with 1 or 2 siphonoglyphs. Septa, both complete and incomplete, often in multiples of 6. Examples: *Anemonia, Actinia, Taelia, Metridium, Calliactis, Edwardsia, Peachia, Epiactis, Stoichactis*.

Order 4. Scleractinia (= Madreporaria). True corals, solitary, or more commonly, colonial anthozoans with a compact calcareous exoskeleton. Radial sclerosepta in hexamerous cycles. No siphonoglyph. Examples: *Fungia, Porites, Acropora, Caryophyllia, Balanophyllia, Meandrina, Siderastrea*.

PHYLUM CTENOPHORA

Biradially symmetrical marine animals, never attached and never colonial. Mesogloea extensive and contains amoebocytes, connective-tissue fibres and muscle fibres. Characterized by eight meridional rows of ciliary plates (comb plates). Apical sense organ. Lack nematocysts.

Class Tentaculata

With two tentacles arising from the body.

Order 1. Cydippida. Rounded or oval. Tentacles retractile into sheaths. Gastrovascular canals end blindly. Examples: *Pleurobrachia, Hormiphora.*

Order 2. Lobata. Two large oral lobes. Tentacles without sheaths. Oral ends of gastrovascular canals anastomose. Examples: *Bolinopsis, Mnemiopsis, Leucothea.*

Order 3. Cestida. Flattened in the tentacular plane. 4 of the comb rows are rudimentary. Tentacles have sheaths, but are reduced. Two small rows of tentacles along oral margin. Examples: *Cestum, Velamen.*

Order 4. Platyctenia. Body flattened in oral-aboral plane. Two tentacles with sheaths. Comb rows may be present only in larvae. Examples: *Ctenoplana, Coeloplana, Tjalfiella.*

Class Nuda

Without tentacles.

Order 1. Beroida. Conical form. Gastrovascular canals with many side-branches. Example: *Beroë.*

5 Classification of Acoelomates

Metazoans with a mouth and gut but generally no anus (the Cestoda, which are endoparasites, lack a digestive system altogether). Body composed of three tissue layers: an outer epidermis, an epithelium lining the digestive cavity when one is present, and between them a true mesoderm in which the gonads are formed. Free-living worms have no cuticle investing the epidermis, parasitic representatives often have a cuticle and may suffer a reduction or loss of the epidermis. Locomotion by cilia or by muscular contraction.

Summary Classification of the Acoelomates

Phylum Platyhelminthes
   Class 1 Turbellaria
        Order 1. Acoela
        Order 2. Rhabdocoela
        Order 3. Alloeocoela
        Order 4. Tricladida
        Order 5. Polycladida
   Class 2 Temnocephaloidea
        Order 1. Temnocephalidea
   Class 3 Monogenea
     Sub-class Monopisthocotylea
        Order 1. Capsaloidea
        Order 2. Gyrodactyloidea
        Order 3. Acanthocotyloidea
        Order 4. Protogyrodactyloidea
        Order 5. Udonelloidea
     Sub-class Polyopisthocotylea
        Order 1. Chimaericoloidea
        Order 2. Diclidophoroidea
        Order 3. Diclybothrioidea
        Order 4. Polystomatoidea

Class 4 Cestodaria
    Order 1. Gyrocotyloidea
    Order 2. Amphilinoidea
Class 5 Cestoda
    Order 1. Pseudophyllidea
    Order 2. Haplobothrioidea
    Order 3. Tetrarhynchoidea (= Trypanorhyncha)
    Order 4. Diphyllidea
    Order 5. Tetraphyllidea
    Order 6. Lecanicephaloidea
    Order 7. Tetrabothrioidea
    Order 8. Proteocephaloidea
    Order 9. Nippotaeniidea
    Order 10. Cyclophyllidea (= Taeniidea)
Class 6 Trematoda
  Sub-class Aspidogastrea (= Aspidobothria)
  Sub-class Digenea
Phylum Rhynchocoela (= Nemertini)
Class 1 Anopla
    Order 1. Palaeonemertini
    Order 2. Heteronemertini
Class 2 Enopla
    Order 1. Hoplonemertini
    Order 2. Bdellomorpha

PHYLUM PLATYHELMINTHES

Bilaterally symmetrical and dorsoventrally flattened worms. Lack a coelom, anus, and circulatory and respiratory systems.

Classification of the Platyhelminthes, particularly of the parasitic forms, presents many difficulties and treatment of them varies widely in different works. The system used here is based on that outlined in Rothschild (1961) *A Classification of Living Animals* (Longmans, London); the chief variations are indicated in appropriate sections below.

Class Turbellaria

Free-living flatworms with a cellular or syncytial epidermis which is usually ciliated at least in part and lacks a cuticle. There are generally secreted rods (rhabdoids) in epidermal glands or cells. These may be extruded to the exterior, but their function is uncertain. Except in the Acoela in which the gut is a syncytium, the digestive cavity is a blind, generally lobed, sac.

Order 1. Acoela. Small worms with a mouth, with or without a pharynx, but no intestine. Endoderm forms a solid syncytial mass. No protonephridia, oviducts, yolk glands, or definitely delimited gonads. Examples: *Convoluta*, *Aphanostoma*.

Order 2. Rhabdocoela. With a complete digestive system, intestine sac-like without diverticula. Generally with protonephridia and oviducts; few, compact gonads. Nervous system generally with two main longitudinal trunks. Examples: *Microstomum*, *Gnathostomula*, *Macrostomum*, *Stenostomum*.

Order 3. Alloeocoela. With a pharynx. Intestine may have short diverticula. Paired protonephridia, often with more than one main branch and nephridiopore. Testes usually numerous, penis papilla usually present. Nervous system with 3 or 4 pairs of longitudinal trunks. Examples: *Prorhynchus*, *Monocelis*.

Order 4. Tricladida. Elongated worms (sometimes very much so). Pharynx generally directed backwards. Intestine in 3 branches, one forwards, two back, all 3 with numerous diverticula. One pair of ovaries, 2 or more testes, a penis papilla, and a single gonopore. Examples: *Planaria*, *Rhynchodemus*, *Procerodes*, *Dugesia*, *Dendrocostum*.

Order 5. Polycladida. Large Turbellaria, generally of a broad, flattened form but sometimes elongated. Pharynx opens into intestine with numerous branches radiating to the periphery. Nervous system with numerous radiating nerve cords. Ovaries and testes numerous and scattered. Male gonopore sometimes separate from the female pore and then anterior to it. Examples: *Leptoplana*, *Thysanozoon*.

Class Temnocephaloidea

Ectocommensals living on the body or in the branchial chamber of freshwater crustaceans, more rarely on freshwater chelonians, and exceptionally in the mantle cavity of freshwater molluscs. External ciliation much reduced or lacking. Anterior end of the body is extended to form 2—12 mobile tentacles. 1—2 pedunculate, adhesive ventral suckers at the posterior end of the body.

Systematic treatment of the Temnocephaloidea varies widely and in

older works they may be found included in the Rhabdocoela or as a separate order in the Turbellaria.
Order 1. Temnocephalidea. Example: *Temnocephala.*

Class Monogenea

Ectoparasites of fish or, more rarely, of amphibians and chelonians; exceptionally found in the cloaca or coelom. Life cycle includes only a single host. Oral sucker weakly developed or, more commonly, absent; flanked by a pair of suckers or adhesive glands. A posterior attachment organ (opisthaptor) is usually armed with spines, hooks, etc. One pair of dorsal, anterior excretory pores. Hermaphrodite. No proper epidermis; body covered by cuticle.

Sub-class Monopisthocotylea

Oral sucker and anterior end also provided with 2 muscular suckers or two glandular adhesive organs. A single opisthaptor. 1—2 testes, no genito-intestinal canal.
Order 1. Capsaloidea. Oviparous monopisthocotyleans with testes anterior to ovary. No vitelline glands. Examples: *Capsala, Tristoma, Dionchus, Dactylogyrus.*
Order 2. Gyrodactyloidea. Viviparous monopisthocotyleans with 1 testis anterior to ovary. No vagina or vitelline gland. A pair of cephalic glands; opisthaptor with 16 marginal spines and usually a pair of central hooks. Example: *Gyrodactylus.*
Order 3. Acanthocotyloidea. One or more testes; vagina present. A pair of cephalic glands. Opisthaptor small, with 16 marginal spines, generally supplemented by a large muscular disc in front of it, covered with concentric rings of radially disposed spines. Example: *Acanthocotyle.*
Order 4. Protogyrodactyloidea. With a single testis. No vagina, but a vitelline gland in 2 or 3 groups with a vitello-intestinal canal. Cephalic glands and opisthaptor with about 12 marginal spines and 2 central hooks linked together by a cross-piece. Example: *Protogyrodactylus.*
Order 5. Udonelloidea. More or less cylindrical monopisthocotyleans with 1 testis but no vagina. A pair of cephalic glands, opisthaptor unarmed. Example: *Udonella* (only genus).

Sub-class Polyopisthocotylea

Only rarely possess an oral sucker; funnel-shaped mouth with an eversible adhesive organ on either side of it. Posterior end with numerous attachment structures (suckers, hooks, etc.). Testes numerous. A genito-intestinal canal.

Order 1. Chimaericoloidea. Posterior end of body drawn out into a long peduncle bearing the opisthaptor composed of 4 pairs of pedunculate fixation organs in the form of a pincer. Example: *Chimaericola*.

Order 2. Diclidophoroidea. Pit enclosing a sucker on either side of the mouth. Opisthaptor with a variable number of sclerotized pincers, sizes vary greatly and sometimes not symmetrically arranged. 1–2 pairs of small hooks often carried on a small process at the posterior end of the opisthaptor. Examples: *Cyclobothrium, Diplozoon, Discocotyle, Gastrocotyle, Hexostoma, Mazocraes, Microcotyle*.

Order 3. Diclybothrioidea. With a terminal sucker surrounding mouth. Opisthaptor with 3 pairs of fixation organs in the form of suckers each enclosing a semicircular sclerotized plate terminating in a point. Opisthaptor sometimes with a posterior prolongation bearing 2 suckers and a pair of small hooks, or 3 pairs of spines. Examples: *Diclybothrium, Hexabothrium*.

Order 4. Polystomatoidea. Opisthaptor with 1 or 3 pairs of large suckers and a pair of large spines, though latter sometimes missing. A variable number of the 16 hooks of the larval opisthaptor persist in the adult. Ecto- and endoparasites of amphibians and reptiles. Examples: *Polystoma, Diplorchis*.

Class Cestodaria
Endoparasites in intestine and body cavity of fish or, exceptionally, chelonians. Body not segmented though sometimes with superficial transverse folds, and generally foliaceous. Lack a scolex and have a single set of reproductive organs.

Sometimes included, as a sub-class, in the Cestoda with the remaining cestodes constituting a separate sub-class Eucestoda.

Order 1. Gyrocotyloidea. Anterior end of body with a terminal sucker. Posterior end drawn out and terminating in a funnel with a frilly mouth or sphincter muscle. Male and uterus pores ventral, vaginal pore dorsal, all at the anterior end of the body. Example: *Gyrocotyle*.

Order 2. Amphilinoidea. Anterior end with protrusible proboscis and frontal glands. Uterus traverses length of body 3 times and opens at anterior end; male and vaginal openings side by side and posterior. Example: *Amphilina*.

Class Cestoda

Endoparasitic tapeworms lacking epidermis and external cilia; body covered by a cuticle. An anterior attachment organ (scolex) with the body usually, though not invariably, divided into a number (sometimes many) segments. No mouth or digestive system. Each segment with more than one set of hermaphroditic reproductive systems. Life cycle complicated, generally involving two or more hosts. Adults live in the intestine of vertebrates.

Systematic treatment of cestodes varies, sometimes with the inclusion of the Cestodaria as a sub-class, with the following orders constituting the sub-class Eucestoda. There is also some variation in the number of small minor orders recognized.

Order 1. Pseudophyllidea. Segmented or unsegmented tapeworms, scolex with 2—6 shallow suckers (bothria), or sometimes without adhesive organs. Yolk glands numerous. Sometimes separated from other cestodes in a sub-class Didesmida, the rest constituting the Tetradesmida. Examples: *Bothriocephalus, Schistocephalus, Diphyllobothrium, Ligula.*

Order 2. Haplobothrioidea. Scolex armed with 4 protrusible proboscides with hooks at their base. Primary strobila fragments into secondary strobilae which have a peculiar pseudoscolex and have an independent life in the host (*Amiá*). Sometimes included in Pseudophyllidea. Example: *Haplobothrium* (only genus).

Order 3. Tetrarhynchoidea (= Trypanorhyncha). Segmented tapeworms, scolex with 4 suckers (bothria) and 4 prostrusible, spiny proboscides. Examples: *Tentacularia, Tetrarhynchus, Hepatoxylon.*

Order 4. Diphyllidea. Segmented tapeworms with up to 20 segments, scolex with 2 suckers (bothria) and a spiny scolex stalk. Example: *Echinobothrium.*

Order 5. Tetraphyllidea. Segmented tapeworms, scolex with 4 suckers (bothria) often armed with hooks. Example: *Phyllobothrium, Acanthobothrium.*

Order 6. Lecanicephaloidea. Segmented tapeworms, scolex with 4 strong muscular suckers (acetabula). Numerous yolk glands. Sometimes included in Tetraphyllidea. Example: *Lecanicephalum.*

Order 7. Tetrabothrioidea. Segmented tapeworms, scolex usually with 4 suckers. Yolk gland compact and anterior to ovary. Functional dorsal uterine pore and gonopores all on one side of strobila. Sometimes included in Cyclophyllidea. Example: *Tetrabothrium.*

Order 8. Proteocephaloidea. Segmented tapeworms, scolex with 4 muscular suckers (acetabula) and an apical adhesive organ (glandular or muscular). Example: *Proteocephalus.*

Order 9. Nippotaeniidea. Segmented tapeworms, scolex with apical sucker but no other adhesive organs. Single yolk gland. Excretory system of numerous longitudinal vessels. Example: *Nippotaenia* (only genus).

Order 10. Cyclophyllidea (= Taeniidea). Segmented tapeworms. Scolex with 4 suckers (acetabula) and apical hooks. Single, compact yolk gland. Excretory system of 4 longitudinal canals. Examples: *Taenia, Dipylidium, Echinococcus.*

Class Trematoda

Endoparasites with a complicated life history with successive larval stages in intermediate hosts. Body unsegmented, lacking epidermis and cilia, covered by a cuticle. Mouth almost always surrounded by a sucker and leads into a complete digestive tract ending in two caeca, sometimes ramifying. Generally a ventral attachment organ in the form of a sucker or adhesive disc. Single posterior excretory pore. Hermaphrodites, though with a tendency to dioecy which may sometimes become complete.

A superficial resemblance between the Monogenea with a single host, and Digenea with multiple hosts, led to them being placed in the same class Trematoda. Although this classification persists in textbooks and is used by some parasitologists, evidence is against it and the Monogenea are commonly separated as an independent class, leaving the Aspidogastrea and Digenea as two groups of trematodes. Treatment of these varies: at one extreme Digenea and Aspidogastrea constitute sub-classes, at the other Monogenea, Digenea and Aspidobranchia are three orders, of the class Trematoda.

Sub-class Aspidogastrea (= Aspidobothria)

No oral sucker or anterior adhesive organs. A very large ventral sucker, subdivided into compartments, or a ventral row of suckers. No hooks. Single posterior excretory pore. Endoparasites with simple life cycle. Examples: *Aspidogaster, Macraspis.*

Sub-class Digenea

Usually with two suckers: an anterior, circum-oral, and a ventral sucker; the former may be lacking but the latter very rarely so. No hooks. Single posterior excretory pore. Uterus a long tube with many capsules. Endo-parasites in natural cavities of the body. Life history complicated with larval stages in one or more intermediate hosts.

Sub-division of the Digenea presents difficulties. There are two classifications, one based on adult structure, the other on the cercaria larva. It is not yet possible to reconcile the two and both have unsatis-factory features. Examples: *Diplostomum, Schistosoma (= Bilharzia)*, *Fasciola, Dicrocoelium, Bucephalus, Clonorchis, Echinostoma.*

PHYLUM RHYNCHOCOELA (= NEMERTINI)

Acoelomates with anus and blood vascular system. Eversible proboscis in a tubular cavity (the rhynchocoel) lying dorsal to the gut.

Class Anopla
Mouth posterior to brain. Nervous system beneath the epidermis or in the body-wall muscle layers. Proboscis unarmed.
Order 1. Palaeonemertini. Body-wall musculature of two or three layers. In the latter case the innermost layer of circular muscles. Example: *Tubulanus.*
Order 2. Heteronemertini. Body-wall musculature of three layers, the innermost longitudinal. Examples: *Lineus, Cerebratulus.*

Class Enopla
Mouth anterior to brain. Nervous system internal to the musculature. Proboscis may be armed.
Order 1. Hoplonemertini. Proboscis armed with one or more stylets. Intestine straight with paired lateral diverticula. Examples: *Amphi-porus, Geonemertes.*
Order 2. Bdellomorpha. Proboscis unarmed. Intestine sinuous and without diverticula. Posterior adhesive disc. Parasitic. Example: *Mala-cobdella.*

6 Classification of Pseudocoelomates

Metazoans in which the body cavity between the gut and body wall is not a true coelom, but is regarded as a persistent blastocoel, though the evidence for this is not always available. Mesodermal derivatives are confined to the body wall and the gut has no muscular coat. The pseudocoel lacks an epithelial lining. (Priapulids, with a muscle coat around the gut but no epithelial lining of the body cavity, are rather uncomfortably accomodated among the pseudocoelomate phyla, possibly incorrectly). No blood vascular system. The first five phyla are closely related to one-another and are commonly regarded as classes of a single phylum Aschelminthes.

Summary Classification of the Pseudocoelomates

Phylum Rotifera
Phylum Gastrotricha
Phylum Kinorhyncha
Phylum Nematoda
   Class 1 Secernentia
   Class 2 Adenophora
Phylum Nematomorpha
Phylum Acanthocephala
Phylum Endoprocta
Phylum Priapulida
All pseudocoelomate phyla, with the exception of the Nematoda, are relatively small and classes are not usually recognized in them.

PHYLUM ROTIFERA

Aquatic, microscopic pseudocoelomates with the anterior end modified into a characteristic ciliary organ (the corona) or a funnel derived from a corona. Metachronal beating of the cilia forming the corona gives the impression of a rotating wheel. Characteristic muscular pharynx with an internal masticatory apparatus composed of a number of hardened (sclerotized) pieces. Body-wall musculature not continuous but in a series of bands. The excretory organs are protonephridia.

Order 1. Seisonidea. Aberrant rotifers epizootic on marine crustaceans. Body very elongated with a long neck piece. Corona only slightly developed. Gonads paired, ovaries lack a yolk gland (vitellarium). Males similar to females. Example: *Seison*.

Order 2. Bdelloidea. Swimming and creeping freshwater forms. Corona with two ciliary discs. Body more or less elongated and made up of a number (about 16) of cylindrical circular joints. Two ovaries with vitellaria. Males unknown, eggs develop parthenogenetically into females. Examples: *Hastotrocha, Rotaria*.

Order 3. Monogononta. Sessile and swimming forms, marine and freshwater. Male more or less reduced, with a single testis and a cirrus or penis. Female with a single ovary and vitellarium. Examples: *Notommata, Testudinella, Collotheca*.

PHYLUM GASTROTRICHA

Microscopic marine and freshwater pseudocoelomates. Body more or less elongated and with cuticular spines, scales, bristles etc. and generally with one or more pairs of adhesive tubes. Locomotion by ventral cilia, but body ciliation limited to certain areas. Nervous system with an anterior ganglionic mass on either side of the pharynx, connected by a dorsal commissure, and with a pair of lateral nerve trunks, running the length of the body. Pharynx structure almost identical with that of nematodes.

Order 1. Macrodasyoidea. Exclusively marine forms with a long, worm-like body equipped with anterior, lateral and posterior adhesive tubes. Locomotion by cilia or by leech-like looping, adhering by the anterior and posterior adhesive tubes in turn. Without protonephridia. Examples: *Macrodasys, Urodasys, Cephalodasys*.

Order 2. Chaetonotoidea. Mostly freshwater forms. With a fusiform body. Adhesive tubes limited to one or two pairs at the posterior end. With protonephridia. Most exist only as parthenogenetic females. Examples: *Chaetonotus, Neodasys, Lepidodermella*.

PHYLUM KINORHYNCHA

Microscopic marine pseudocoelomates, body spiny and made up of 13—14 jointed sections. Head armed with circles of spines and can be withdrawn into the second or third section. Lack cilia. Body flattened below, arched above, with more or less parallel sides. Nervous system with a circum-oral or circum-pharyngeal ring and a mid-ventral nerve strand with ganglia in each body section. Examples: *Echinoderes, Centroderes*.

PHYLUM NEMATODA

Vermiform, unsegmented, more or less cylindrical pseudocoelomates with a terminal mouth and an almost terminal anus (usually a short post-anal caudal region) and a straight non-muscular gut. Tri-radiate muscular pharynx. Body covered by a thick, non-chitinous cuticle which may be strongly annulated giving a false impression of segmentation. No flame cells. Four thickened epidermal strands running the length of the body. Sexes separate, but some species are protandric hermaphrodites, and some exist only as parthenogenetic females. Each species with a fixed, characteristic number of cells except in the gonads (eutely). Free-living and parasites of plants and animals. Occur in all environments and in all plants and animals. Free-living species are small (0.5–3.0 mm long), parasitic species living in animals are often much longer.

Taxonomy (especially of higher taxa) is chaotic and several different systems are in use.

Class Secernentea (= Phasmidia)

Special secretory and sense organs (phasmids) on sides of post-anal region (Fig. 5). No caudal glands. Sensory papillae low. Cervical papillae usually present. Sensory pits (amphids) (Fig. 5) open on lateral lips. Excretory system with coiled collecting ducts. Male often with caudal 'wings'.

Order 1. Tylenchida. Cuticle annulated. Anterior part of gut with a hollow, piercing spear. Oesophagus with posterior and median bulb, the latter with heavy plates. Lateral field with longitudinal incisions. 14 small head papillae. 1 or 2 gonads. Important plant parasites, but many free-living. Examples: *Hoplolaimus, Radopholus, Aphelenchoides, Anguina* (= *Tylenchus*).

Order 2. Rhabditida. Oesophagus with large basal bulb, anterior part of oesophagus usually swollen and separated from bulb by constriction. Excretory system symmetrical, H- or U-shaped. Many free-living, often in dung, some semiparasitic, some parasitic. Parasitic forms generally have a resistant 3rd larval stage. A very large order (sometimes split

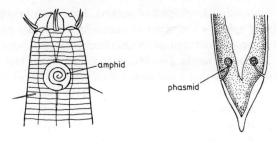

FIGURE 5   Amphids and phasmids of nematodes (after Hyman).

into 5 orders: Rhabditoidea, Tylenchoidea, Strongyloidea, Oxyuroidea, Ascaroidea). Examples: *Rhabditis, Panagrellus, Strongylus, Nippostrongylus, Haemonchus, Ascaris.*

Order 3. **Spirurida.** Oesophagus divided into anterior muscular and posterior glandular regions. Males with well developed caudal wings and long spiral tails. Mostly parasites of man and large mammals. Larvae with an invertebrate intermediate host. Examples: *Litomosoides, Wucheria, Onchocerca, Gongylonema, Thelazia.*

Order 4. **Camallanida.** Long, thin worms; mouth often with lateral jaws. Parasites of vertebrate connective tissues; larvae usually in copepods. Example: *Dracunculus.*

Class Adenophora (= Aphasmidia)

Lack phasmids. Often with caudal glands. Sensory papillae generally bristle-like. Amphids (Fig. 5) open far back on head capsule, usually near the suture. Excretory system without collecting ducts. No caudal wings or papillae. Oesophagus cylindrical. Some have buccal spears, but very different from Secernentea.

Order 1. **Monhysterida.** No buccal spear. Cuticle often with bristles; head with bristles. Mouth often armed with teeth. Free-living, often predatory. Examples: *Tripyla, Monhystera, Plectus, Mononchus.*

Order 2. **Dorylaimida.** With a spear (tubular, like a hypodermic needle). Cuticle smooth, without bristles. Teeth are often present in forms with the spear reduced. Predators and plant parasites. Examples: *Dorylaimus, Longidorus, Trichodorus.*

Order 3. **Chromadorida.** With spiral amphids. Cuticle usually annulated and with numerous punctations, knobs and bristles. Buccal capsule often armed with teeth. Pharynx has a posterior bulb. Examples: *Plectus, Chromodora.*

Order 4. Enoplida. Amphids like pockets, head with bristles. Male has prominent spicules and often an accessory copulatory apparatus. Aquatic, freshwater and marine. Examples: *Enoplus, Thoracostoma*.

Order 5. Trichosyringida (= Mermithoidea). Long (15–20 cm), slender worms. Smooth cuticle with cuticular cross fibres visible by transparency. Larvae with a stylet. In the adult the intestine is replaced by a trophosome (equivalent to a fat body). Egg-bearing females with red spot at back of head. Larvae parasitic in arthropods. Adults free-living in soil. Examples: *Mermis, Agamermis*.

Order 6. Trichurata. Oesophagus very long and capillary-like, with numerous large oesophageal glands (the whole oesophageal structure is termed a 'stichosome'). Anterior end, containing the oesophagus, much narrower than the rest of the body. Anus terminal. Parasites of vertebrates. Examples: *Trichuris, Capillaria, Trichinella*.

Order 7. Dioctophymatida. Caudal end of male with muscular, bell-shaped bursa. No lips, but the mouth surrounded by one or two circles of papillae. Small group of large nematodes, parasitic in birds and mammals (often in the kidneys or peritoneum), intermediate hosts are probably fishes. Example: *Dioctophyma*.

PHYLUM NEMATOMORPHA

Exceedingly long, thin pseudocoelomate worms, parasitic in arthropods as juveniles, free-living, generally in soil or freshwater, as adults. Digestive system more or less degenerate at anterior and posterior ends. Body-wall musculature of longitudinal muscle cells only. Sexes separate, both with a cloaca. Single mid-ventral nerve cord.

Order 1. Gordioidea. Pseudocoel occluded with mesenchymal tissue. Nerve cord beneath the epidermis and lying in the pseudocoel. Paired gonads. Adults in soil or freshwater. Examples: *Gordius, Chordodes*.

Order 2. Nectonematoidea. Marine pelagic form with numerous swimming bristles on body. Pseudocoel not occluded. Ventral gonad. Example: *Nectonema* (only genus).

PHYLUM ACANTHOCEPHALA

Intestinal parasites of vertebrates, with a long, more or less cylindrical body and a characteristic eversible proboscis at the anterior end, which is covered with recurved spines and used as an attachment organ. No digestive system or blood vascular system. Epidermis syncytial and containing a system of lacunae. Nervous system consists of an anterior ganglion and a pair of lateral nerve cords. Main body cavity a pseudocoel and contains the reproductive system, but little else. Sexes separate. Development via a larva which spends some time in an arthropodan intermediate host.

Order 1. Archiacanthocephala. Main lacunar channels in epidermis mid-dorsal and mid-ventral. Proboscis spines arranged concentrically. Generally with protonephridia. Parasites of terrestrial hosts. Example: *Macracanthorhynchus*.

Order 2. Palaeacanthocephala. Main lacunar channels in epidermis lateral. Proboscis spines arranged in alternating radial rows. Lack protonephridia. Mostly in aquatic hosts. Examples: *Acanthocephala, Echinorhynchus*.

Order 3. Eoacanthocephala. Main lacunar channels in epidermis mid-dorsal and mid-ventral. Proboscis spines arranged in radial rows. Lack protonephridia. Parasites of aquatic hosts. Example: *Neoechinorhynchus*.

PHYLUM ENDOPROCTA

Solitary or colonial, stalked or sessile pseudocoelomates. The body is divided into a rounded mass containing the viscera (the calyx) and a stalk by which the animal is attached to the substratum. A circle of tentacles arises from the free edge of the calyx, and both mouth and anus open to the exterior within this tentacular ring. The gut is U-shaped. Three families; no orders or classes recognized. Examples: *Loxosoma, Pedicellina, Urnatella*.

PHYLUM PRIAPULIDA

Marine, burrowing pseudocoelomates. Body cylindrical and superficially annulated and warty. The anterior is part armed with several concentric series of circum-oral spines, and can be introverted into the remainder of the body. Gut straight, and mouth and anus terminal. A pair of urinogenital organs with gonads and protonephridia, opening to the exterior at two urinogenital pores beside the anus. *Priapulus* with one or two caudal appendages like bunches of grapes. The main body cavity is regarded as a pseudocoel, but this is not certain. Probably only three species in two genera: *Priapulus, Halicryptus*.

7 Classification of Protostome Coelomates

Metazoans with a true *coelom* (i.e. a secondary body cavity lying completely within the mesoderm), or derived from animals with a true coelom. A patent coelom exists only in Echiura, Sipuncula, and Annelida, is of a doubtful nature and at best rudimentary in Mollusca, and has become confluent with the blood vascular system to form a *haemocoel* in the remaining phyla. Protostome characteristics are largely embryological and are most clearly observed in members of phyla which retain a patent coelom, viz: cleavage of the egg is spiral, the blastopore becomes the mouth, or mouth and anus, the coelom is generally formed by schizocoely, typical form of the larvae is a trochophore with ciliary bands encircling the body.

Summary Classification of Protostome Coelomates

Phylum Sipuncula
Phylum Echiura
Phylum Mollusca
    Class 1 Monoplacophora
    Class 2 Amphineura
      Sub-class 1 Polyplacophora
      Sub-class 2 Aplacophora
    Class 3 Gastropoda
      Sub-class 1 Prosobranchia
      Sub-class 2 Opisthobranchia
      Sub-class 3 Pulmonata
    Class 4 Scaphopoda
    Class 5 Bivalvia
      Sub-class 1 Protobranchia
      Sub-class 2 Lamellibranchia
    Class 6 Cephalopoda

Sub-class 1 Nautiloidea
Sub-class 2 Ammonoidea
Sub-class 3 Coleoidea
Phylum Annelida
    Class 1 Polychaeta
    Class 2 Myzostomaria
    Class 3 Clitellata
        Sub-class 1 Oligochaeta
        Sub-class 2 Hirudinea
Phylum Tardigrada
Phylum Pentastomida
Phylum Onychophora
Phylum Arthropoda
  Sub-phylum 1 Trilobitomorpha
  Sub-phylum 2 Chelicerata
    Class 1 Merostomata
    Class 2 Arachnida
    Class 3 Pycnogonida
  Sub-phylum 3 Mandibulata
    Class 1 Crustacea
        Sub-class 1 Cephalocarida
        Sub-class 2 Branchiopoda
        Sub-class 3 Ostracoda
        Sub-class 4 Copepoda
        Sub-class 5 Mystacocarida
        Sub-class 6 Branchiura
        Sub-class 7 Cirripedia
        Sub-class 8 Malacostraca
    Class 2 Chilopoda
        Sub-class 1 Epimorpha
        Sub-class 2 Anamorpha
    Class 3 Diplopoda
        Sub-class 1 Pselaphognatha
        Sub-class 2 Chilognatha
    Class 4 Symphyla
    Class 5 Pauropoda
    Class 6 Insecta
        Sub-class 1 Apterygota
        Sub-class 2 Pterygota

PHYLUM SIPUNCULA

Unsegmented coelomate worms. The whole anterior end of the body, with the mouth surrounded by a circle of tentacles at its tip, can be introverted and used as a proboscis. Gut coiled upon itself so that the anus is near the anterior end of the body. Coelom spacious. Exclusively marine, chiefly sedentary animals living in crevices in rocks, etc., or in burrows in the substratum. The phylum contains 12 genera. Other taxa classes, orders, etc.) are not used. Examples: *Sipunculus, Phascolion, Golfingia, Dendrostomum.*

PHYLUM ECHIURA

Unsegmented coelomate worms, although there are rather questionable transitory signs of segmentation of the mesoderm during embryonic development. Extensible proboscis at anterior end, not eversible, with a ciliated groove and used as a food collecting organ. The mouth lies at the base of the proboscis. Gut coiled but the anus is posterior and terminal. Coelom spacious and undivided. Most echiurids have a few specialized chaetae (copulatory spines, etc.) which are formed in precisely the same way as those of annelids. Exclusively marine, sedentary animals living in burrows in a sandy or muddy substratum. More than 200 species in about 30 genera distributed between five families: Echiuridae, Thalassemidae, Borelliidae, Urechidae and Ikedaidae. The first three families are sometimes grouped in an order Echiurida, with the Urechidae and Ikedaidae constituting separate orders Xenopneusta and Heteromyota, respectively, but classification above the level of family is not regarded as satisfactory and the orders are not generally used. Examples: *Echiurus, Thalassema, Bonellia, Urechis, Ikeda.*

PHYLUM MOLLUSCA

Generally regarded as coelomate (in which case the coelom is represented only by the pericardial cavity) and occasionally as segmented although the evidence for this is weak. The body has a basic bilateral symmetry and theoretically is made up of: (i) a central concentration of the viscera (the 'visceral mass') covered by a mantle ('pallium') which grows out peripherally, (2) a head and (3) a foot. In fact, the body form is extremely plastic and subject to the widest modification so that molluscs have no recognizable stereotyped body pattern as the arthropods and annelids have. Most molluscs have branchiae of a characteristic type ('ctenidia') but other types of gill may exceptionally be developed. Development frequently includes a veliger larval stage which is similar to a trochophore except that it has a special locomotory and food-collecting organ, the velum, developed from the pre-oral ciliary band.

The following classification is that given in Morton and Yonge (*Physiology of Mollusca* (eds. K. M. Wilbur & C. M. Yonge) 1, 1—58. Academic Press, New York and London, 1964).

Class Monoplacophora
Almost bilaterally symmetrical molluscs with terminal mouth and anus and a single limpet-like shell. The foot is a ventral muscular disc. Shallow mantle cavity around the foot encloses 5 pairs of branchiae with filaments on one side (uniseriate). Dorso-lateral and ventral cavities are described as coelomic although they do not correspond with the coelom of other molluscs. Two pairs of auricles and two pairs of gonads discharging through two of the six pairs of renal organs. Mostly extinct, the few modern forms are known only from the seabed at some depth and are members of the order Tryblidioidea. Example: *Neopilina*.

Class Amphineura
Elongated, bilaterally symmetrical molluscs with terminal mouth and anus. Mantle covers the dorsal surface and sides of the body. Heart posterior and dorsal with a ventricle and lateral auricles.

The two constituent sub-classes are raised to the status of separate classes by some authors.

Sub-class Polyplacophora
Flattened and with a broad ventral foot, covered dorsally by 8 transverse shell plates bordered by a girdle with spicules or scales. Mantle cavity around the foot encloses numerous ctenidia in practice dividing the mantle cavity into anterior outer inhalent and posterior inner exhalent compartments. Characteristically intertidal but a few found in deep water. Classification is based chiefly on features of the shell plates.

Order 1. Lepidopleurida. Primitive chitons with simple digestive and renal systems. Ctenidia confined to the posterior end of the mantle, flanking the anus. Example: *Lepidopleurus*.

Order 2. Chitonida. Digestive and renal systems advanced. Ctenidia in the greater part, if not all the mantle groove and generally not close to the anus. Examples: *Chiton, Tonicella, Cryptochiton, Mopalia, Acanthochitona.*

Sub-class Aplacophora

Aberrant, worm-like, elongated amphineurans without a shell. The mantle completely encloses the body, except in some for a longitudinal ventral groove. Mantle studded with numerous calcified spicules.

Order 1. Neomeniomorpha. Median longitudinal ventral groove in mantle containing a linear vestige of the foot. Living on corals or hydroids. Examples: *Neomenia, Proneomenia.*

Order 2. Chaetodermomorpha. No ventral groove in the mantle and no vestiges of a foot. Live in deep water feeding on deposits in the ooze. Example: *Chaetoderma.*

Class Gastropoda

Asymmetrical molluscs with a single shell, coiled in a helical spiral, which is formed in the young stages and generally persists in the adult. Well developed head and, at least primitively, a broad flattened foot. The visceral mass with the mantle has rotated 180° on the foot (torsion) but in a number of gastropods this process is partially reversed (detorsion). Because of the asymmetrical coiling (as distinct from torsion) of the visceral mass, the pallioperitoneal complex (i.e. heart, renal organs, gonads, ctenidia) is usually reduced and one-sided.

Sub-class Prosobranchia

Aquatic and with a spiral shell closed by an operculum. Head with a single pair of tentacles with eyes at the base. Visceral mass retains pronounced torsion and the visceral loop of the nervous system is crossed into a figure-of-eight. Mantle cavity with two ctenidia in some, but usually one (on the post-torsional right) is reduced. A single gonad opening on the right, either through the renal organ (if the left renal organ is suppressed) or through the renal duct (if the left renal organ is retained and functional), in the latter case the genital duct becomes elaborate.

Order 1. Archaeogastropoda. 2 or 1 ctenidia with filaments on both sides (bipectinate) (Fig. 6), but limpets may have secondary gills in the mantle grooves (Patellidae) or lack gills altogether (Lepetidae). Heart with 2 auricles. Right renal organ is the functional kidney. Nervous system not greatly concentrated. Live in marine habitats. Examples: *Acmaea, Haliotis, Patella, Pleurotomaria, Trochus*.

Order 2. Neritacea. Left renal organ enlarged and becomes the functional kidney, right renal organ lost but its duct becomes part of the genital tract. Males with a cephalic penis; females with glandular genital tract. Nervous system not greatly concentrated. Single bipectinate ctenidium. Have invaded freshwater and terrestrial habitats. Example: *Nerita*.

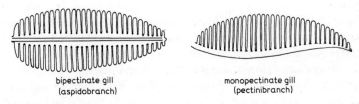

bipectinate gill
(aspidobranch)

monopectinate gill
(pectinibranch)

FIGURE 6   Gastropod gills.

Order 3. Mesogastropoda. Shell opening sometimes with a groove to accommodate a siphon (siphonate). Organs of right side of pallioperitoneal complex lost. Single ctenidium with filaments on one side only (monopectinate, or pectinibranch) (Fig. 6), osphradium well developed and sometimes pectinate. Left renal organ a functional kidney, right renal duct the genital duct which has glandular extensions of the mantle and produces egg capsules or a jelly egg mass; cephalic penis; fertilization internal. Nervous system somewhat concentrated. Generally carnivorous, some with an eversible proboscis. Examples: *Littorina, Strombus, Cypraea, Carinaria, Viviparus, Aporrhais, Crepidula, Charonia, Cassis, Pterotrachea, Bithynia, Hydrobia*.

Order 4. Neogastropoda (= Stenoglossa). Shell always siphonate. Carnivorous, feeding on living and dead animals, and with an eversible proboscis. Osphradium large and bipectinate. Eggs usually develop in an egg capsule and free-swimming veliger larvae are usually suppressed. Examples: *Buccinum, Busycon, Nassa, Ocenebra, Terebra, Conus*.

Sub-class Opisthobranchia

Shell reduced, becoming internal or lost altogether. A tendency to detorsion, with the mantle cavity moving back along the right side and developing a large opening to the exterior; it is lost in some. Visceral loop of the nervous system in various degrees uncrossed and shortened. Gill probably never a true ctenidium. Accompanying detorsion and loss of the shell, the appearance of external bilateral symmetry and a considerable range of morphological forms.

The classification of opisthobranchs is by no means settled. Differences arise chiefly over the number of orders in the sub-class. In older works only four orders are recognized but more recently there has been a tendency to subdivide three of them, giving a total of eight orders. Some authors further subdivide the Acoela into a number of orders.

Order 1. Cephalaspidea (= Bullomorpha). Shell fairly large and mantle cavity well developed, with a single folded gill. The head forms a large shield for burrowing. Parapodia prominent and sometimes finlike. Examples: *Actaeon, Retusa, Bulla, Scaphander*.

Order 2. Anaspidea (= Aplysiomorpha). Shell reduced and internal, mantle cavity a small recess on right side. No head shield; animals crawling or swimming by enlarged parapodia. Examples: *Apylsia, Akera*.

Order 3. Thecosomata. Planktonic pteropods with parapodial fins, and a spirally coiled shell, or a modified nonspiral 'pseudoconch'. Pallial cavity well-developed. Examples: *Limacina, Cavolina, Spiratella*.

Order 4. Gymnosomata. Naked planktonic pteropods with small ventral parapodial fins; externally symmetrical, and fast-swimming. No shell or mantle cavity. Examples: *Clione, Pneumoderma*.

Order 5. Notaspidea (= Pleurobranchomorpha). Shell reduced and internal, without a mantle cavity, but a naked gill overhung by the mantle on the right side. Becoming flattened, slug-like, and externally almost symmetrical. Examples: *Umbraculum, Pleurobranchus*.

Order 6. Acochlidiacea. Very small, interstitial sand-dwelling opisthobranchs, visceral mass marked off as a long hump from the foot, without dorsal appendages, though with spicules. Examples: *Acochlidium, Microhedyle*.

Order 7. Sacoglossa. Herbivorous suctorial opisthobranchs with characteristically modified radula and buccal mass. Running from primitive shelled and spirally coiled forms to slug-like 'nudibranchs'. Examples: *Elysia, Limapontia*.

Order 8. Acoela (= Nudibranchia). Naked, externally almost symmetrical slug-like forms with no mantle cavity or external shell. Dorsal integument has outgrowths such as cerata, or pinnate retractile gills encircling a median anus. Examples: *Onchidoris, Glaucus, Pleurophyllidea, Doris, Eolis*.

Sub-class Pulmonata
Lack a ctenidium, but the mantle cavity is vascularized as a lung and the small mantle opening is contractile. Detorsion is seldom complete, but the nervous system is highly concentrated and the nerve cords are not twisted. The shell and visceral mass are primitively spiral but the former may be reduced and the body assume a slug-like form. Hermaphrodite.

Order 1. Basommatophora. Head with a single pair of non-invaginable tentacles with eyes at base. Most species are aquatic, primitively or by reversion, and may acquire a secondary gill. Examples: *Lymnaea, Planorbis, Ancylus*.

Order 2. Stylommatophora. Two pairs of invaginable tentacles, with eyes on summit of the hinder pair. Terrestrial snails, giving rise by loss of spiral shell to slugs. Examples: *Helix, Testacella, Limax, Arion*.

Class Scaphopoda
Bilaterally symmetrical, with the mantle and shell fused ventrally to form a tapered tube open at each end. Head without eyes but bears paired clusters of food-capturing organs (capitula). The foot is cylindrical and pointed. No ctenidia. The sexes are separate. There are no special genital ducts and fertilization is external. Scaphopods are exclusively marine. Examples: *Cadulus, Dentalium, Siphodentalium, Pulsellum*.

Class Bivalvia
Bilaterally symmetrical and laterally compressed molluscs with a shell consisting of two valves hinged together dorsally by a ligament and with projections from the valves forming hinge teeth. Head rudimentary, foot laterally compressed and without a plantar surface. Ctenidia generally very large and, together with labial palps, used in ciliary feeding as a rule. No radula. Sub-classes are not always recognized.

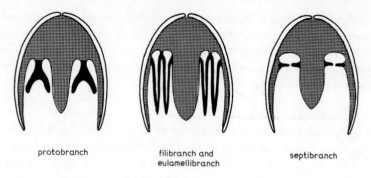

protobranch   filibranch and   septibranch
                     eulamellibranch

FIGURE 7   Bivalves: diagrammatic sections to show gills (ctenidia).

Sub-class Protobranchia

Except in Solemyidae, feeding primarily by means of extensible 'proboscides' from the enlarged labial palps. Ctenidia with flat, non-reflected filaments (Fig. 7) or may be fused and muscular, forming 'pumping' ctenidia. Hypobranchial glands retained. Foot with numerous retractors and opening out to expose a flattened ventral surface. A single order Protobranchia. Examples: *Nucula, Yoldia, Solemya*.

Sub-class Lamellibranchia

Ctenidia are the principal feeding organs and have very long filaments folded back on themselves (reflected) (Fig. 7) so that a row of filaments forms two lamellae which are generally united by lamellar junctions. Filaments usually attached to adjacent ones by ciliary discs (filibranch condition), or by tissue bridges (eulamellibranch condition). Labial palps relatively small compared with those of Protobranchia.

Order 1. Taxodonta. Gill filaments free and without interlamellar junctions. Shell hinge with numerous similar teeth. Edges of the mantle not fused together. Anterior and posterior adductor muscles nearly equal. Examples: *Arca, Glycymeris*.

Order 2. Anisomyaria. Gills usually filibranch with vascular interlamellar junctions. Edges of the mantle sometimes fuse to separate off an exhalent opening but are otherwise free. Anterior adductor muscle reduced and sometimes lost altogether (monomyarian condition) (Fig. 8). This leads to a radical rearrangement of the symmetry of the body. Foot is small and sometimes absent. Frequently attached to the substratum by byssus threads or one shell valve cemented to the substrate.

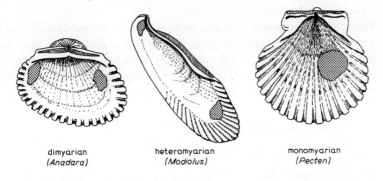

dimyarian	heteromyarian	monomyarian
(Anadara)	(Modiolus)	(Pecten)

FIGURE 8   Position of bivalve adductor muscles (after Hunter and Brown).

Examples: *Modiolus, Mytilus, Ostrea, Pecten, Pinctada, Pinna, Anomia.*

Order 3. Heterodonta. Gills usually eulamellibranch. Mantle edges usually fused at one or more points ventrally and often extended posteriorly into siphons. Adductor muscles equal one-another. Hinge teeth of the shell 'heterodont'. Mostly burrowing forms. Examples: *Astarte, Cardium, Tellina, Macoma, Mactra, Spisula, Tridacna.*

Order 4. Schizodonta. Probably an artificial group which have in common 'schizodont' hinge teeth. The gills are eulamellibranch. Examples: *Trigonia* (Trigoniacea), *Anodonta, Unio* (Unionacea).

Order 5. Adepodonta. Gills eulamellibranch. Mantle edges completely fused together ventrally except where the foot emerges. Mantle extended posteriorly to form long united siphons, sometimes with the gills extending into them. The shell ligament is weak or absent and the shell gapes. Deep, permanent burrowers, sometimes in hard substrates. Examples: *Ensis, Mya, Pholas, Teredo.*

Order 6. Anomalodesmata. Gills eulamellibranch. Mantle edges extensively fused together. No hinge teeth. Foot small. Hermaphrodites with separate male and female genital openings. Examples: *Pandora, Lyonsia, Thracia.*

Order 7. Septibranchia. Gills small and transformed into a transverse, muscular pumping septum, dividing the mantle cavity into an inhalent chamber below and a suprabranchial cavity above (Fig. 7). Mantle edges not extensively fused together. Hinge teeth reduced or lacking. Macrophagous, feeding on animal remains. Sometimes raised to the status of a sub-class. Examples: *Cuspidaria, Poromya.*

Class Cephalopoda

Bilaterally symmetrical molluscs with a circle of prehensile tentacles around the mouth. Ventrally, part of the foot is modified to form a pallial funnel through which the muscular mantle produces a concentrated jet of exhalent water which can be used in locomotion. Sense organs highly developed and the central nervous system large and concentrated.

Sub-class Nautiloidea

Cephalopods with an external, many-chambered shell. Head with numerous retractile tentacles which have no suckers. Funnel composed of two separate folds. Two pairs of ctenidia and renal organs. Eyes without cornea or lens. Formerly very numerous but now represented only by the genus *Nautilus*.

Sub-class Ammonoidea

Totally extinct. Nothing reliably known of the structure of the animals but in general comparable with, and undergoing a similar radiation to the nautiloids.

Sub-class Coleoidea

In living forms the mantle is naked, forms a sac covering the viscera and contains a more or less rudimentary shell. The head always has eight sucker-bearing arms, and there may be an additional pair of longer and retractile arms between the third and fourth short pairs. The funnel is always a closed tube. A single pair of ctenidia and renal organs. Eye with a crystalline lens and closed or open cornea. Ink sac present.

Order 1. Decapoda. Tentacular retractable arms in addition to eight normal arms which are shorter than the body. Suckers pedunculate with horny rings. Internal shell relatively well-developed. Squids and cuttlefish. Examples: *Architeuthis*, *Sepia*, *Loligo*, *Spirula*.

Order 2. Octopoda. Eight uniform arms longer than the body, with non-pedunculate suckers. The mantle encloses the viscera in a rounded muscular sac and there is no internal shell, although the female *Argonauta* has an external 'shell' secreted by the dorsal arms. Examples: *Octopus*, *Argonauta*, *Eledone*.

Order 3. Vampyromorpha. 'Vampire squids' with eight long arms united by a swimming web and two small, retractile, tendril-like arms. Example: *Vampyroteuthis*.

PHYLUM ANNELIDA

Metamerically segmented, coelomate worms with a non-segmental element at the anterior (prostomium) and posterior (pygidium) ends. Alimentary canal with mouth and anus, and generally straight. Metamerism is distinct and generally not substantially modified. Body covered with a thin, flexible cuticle. The taxonomy of the phylum is unsatisfactory; usage varies a good deal and not always consistently. The following classification is that in Clark (in *Chemical Zoology* (eds M. Florkin & B. T. Scheer) 4, 1—68. Academic Press, New York & London, 1969).

Class Polychaeta

Typically, the worms have a pair of lateral, lobed appendages (parapodia) on each segment, bearing numerous bristles or chaetae. Both the parapodia and chaetae are subject to great variation, as is the morphology of the worms so that the typical characteristics may not be immediately apparent. The anterior non-segmental prostomium is often very much reduced or may be obliterated by fusion with the anterior segments. The anterior end generally bearing sensory equipment—eyes, tentacles, palps, antennae, etc., some of which may be modified as food collecting organs. Mostly marine worms; a few in freshwater and half a dozen terrestrial. The non-marine species are nearly all in the family Nereidae.

There is no generally accepted and satisfactory subdivision of the class. The Polychaeta fall into 66 well-defined families but the relationships between them are not clear. Grouping them into two sub-classes Errantia and Sedentaria, though common, is taxonomically indefensible and simply a matter of convenience. There is a fair measure of agreement as to the constitution of the twelve orders although they are not often used. The more important families in each order are indicated in capitals.

Order 1. Amphinomorpha. Segments all more or less alike and bearing well developed parapodia. Prostomium small and drawn back to lie between the anterior segments. Mouth and proboscis ventrally directed; the latter with a chitinized rasping organ used in food collection. Families: Amphinomidae, Euphrosynidae, Spintheridae.

Order 2. Eunicemorpha. Metamerism not modified externally and segments all more or less alike. Prostomium distinct and often bearing tentacles, although these are reduced or lacking in the Lumbrineridae and Arabellidae. Eversible proboscis directed ventrally, not axially, armed with replaceable teeth and jaws. Parapodia may be reduced. Families: EUNICIDAE, Onuphidae, LUMBRINEREIDAE, ARABELLIDAE, Lysaretidae, Dorvilleidae, Histriobdellidae, Ichthyotomidae.

Order 3. Phyllodocemorpha. Metamerism not modified externally and segments all more or less alike. Prostomium distinct and bearing sensory appendages, none of which are modified as food-collecting organs. Parapodia well developed. Eversible proboscis disposed in the longitudinal axis of the body, and generally armed with sclerotized jaws. Most of the typically 'errant' polychaetes are in this order. Families: PHYLLODOCIDAE, Alciopidae, Lopadorrhynchidae, Lacydoniidae, Iospilidae, Pontodoridae, Typhloscolecidae, TOMOPTERIDAE, APHRODITIDAE, POLYNOIDAE, POLYODONTIDAE, SIGALIONIDAE, Pareulepidae, Chrysopetalidae, Palmyridae, Pisionidae, HESIONIDAE, Pilargiidae, SYLLIDAE, NEREIDAE, NEPHTYIDAE, Sphaerodoridae, GLYCERIDAE, GONIADIDAE.

Order 4. Spiomorpha. Body composed of numerous segments, generally all more or less alike. A tendency for the prostomium to fuse with the anterior segments. A pair of long tentacles develop on the posterior margin of the prostomium, but generally shift into a more anterior position in the adult, modified as food-collecting organs. Proboscis directed ventrally, without jaws, and often reduced. Families: ORBINIIDAE, Paraonidae, Apistobranchidae, SPIONIDAE, MAGELONIDAE, Disomidae, Poecilochaetidae, Longosomidae, CHAETOPTERIDAE, CIRRATULIDAE, Ctenodrilidae, Stygocapitellidae, SABELLARIIDAE.

Order 5. Drilomorpha. Prostomium small and lacking appendages of any kind. Proboscis thin-walled and soft. Burrowing, mud-eating worms. Families: CAPITELLIDAE, ARENICOLIDAE, MALDANIDAE, SCALIBREGMIDAE, OPHELIIDAE.

Order 6. Oweniimorpha. Body composed of few, long segments. Prostomium fused with succeeding segment and generally terminating in a frilled circum-oral lobe which is used as a food-collecting organ. Family: OWENIIDAE.

Order 7. Sternaspimorpha. Body globular, with only faint traces of segmentation. No true proboscis, but the anterior part of the body can be introverted. A unique, ventral, posterior chitinous plate. Anal gills. Family: STERNASPIDAE.

Order 8. Flabelligerimorpha. Body cylindrical, tapering at each end, with short segments all more or less alike and covered with papillae. Prostomium and first segment bearing two palps and numerous short gills, all of which can be invaginated. No true proboscis. Families: FLABELLIGERIDAE, Poeobiidae.

Order 9. Terebellomorpha. Sedentary, tubicolous worms living in burrows or tubes. Prostomium fused with anterior segments and indistinguishable. Numerous food collecting tentacles at the anterior end. Segments of more than one kind and posterior part of body often forming a 'tail'. Families: TEREBELLIDAE, TRICHOBRANCHIDAE, PECTINARIIDAE, AMPHARETIDAE.

Order 10. Serpulimorpha. Prostomium fused with anterior segments and indistinguishable. A cone-shaped crown of gills at the anterior end, used for filtering particles from the water. Families: SABELLIDAE, SERPULIDAE.

Order 11. Psammodrilomorpha. Minute worms living as an interstitial fauna. A unique pumping pharyngeal apparatus derived from the longitudinal body-wall muscles. Family: Psammodrilidae.

Order 12. Archiannelida. Small, active worms composed of relatively few segments all more or less alike, living as an interstitial fauna. Parapodia, and generally other signs of segmentation reduced. Extensive body ciliation used for locomotion. Protrusible muscular buccal apparatus in the floor of the buccal cavity used in food collecting (absent in *Polygordius*). Once considered a separate class of annelids but now regarded as polychaetes which have undergone considerable modification and secondary simplification as a consequence of their small size and adaptation to the interstitial habitat. Families: NERILLIDAE, PROTODRILIDAE, DINOPHILIDAE.

Class Myzostomaria

Dorso-ventrally flattened and almost circular in outline. Five pairs of parapodia, but most signs of segmentation lost. Exclusively ectoparasitic on echinoderms, chiefly crinoids. Example: *Myzostomum*.

Class Clitellata

Hermaphrodite annelids with few or no chaetae and no parapodia. Epidermis of a few consecutive segments in the middle of the body glandular and constituting a clitellum which secretes a cocoon for the eggs. The clitellum may be distinguishable only in the breeding season.

The systematic treatment of the class Clitellata is not consistent. Oligochaete systematists generally do not recognize it but regard oligochaetes as representing a separate class. Hirudinean systematists, on the other hand, generally recognize it and regard the Oligochaeta and Hirudinea as constituent sub-classes of the Clitellata. The latter view reflects more clearly the systematic relationships of these annelids.

Sub-class Oligochaeta

Segments all alike externally with a few chaetae arising directly from the body wall. Prostomium small and lacking appendages. Only a few segments contain gonads, the testes in segments anterior to the ovaries. Coelom spacious and subdivided by intersegmental septa. Chiefly terrestrial and freshwater forms, a few marine.

The status of the four orders is not altogether clear and frequently they are ignored. The first three contain small aquatic worms often referred to as 'microdriline' or 'limicoline', the fourth order contains the earthworms which are often referred to as 'megadriline' or 'terricoline'.

Order 1. Prosopora. Male gonopores on the segment that contains the testes, or on the same segment as the last pair of testes. Families: Lumbriculidae, Branchiobdellidae.

Order 2. Plesiopora Plesiothecata. Male gonopores open on the segment immediately behind that containing the testes. Spermatheca in the same regions as gonads. Families: Aeolosomatidae, Naididae, Opisthocystidae, Tubificidae, Phreodrilidae.

Order 3. Plesiopora Prosothecata. Male gonopores open on the segment immediately behind that containing the testes. Spermatheca in a segment a considerable distance anterior to the gonads. Family: Enchytraeidae.

Order 4. Opisthopora. Male gonopores on the segment some distance behind the last pair of testes. Families: Haplotaxidae, Alluroididae, Syngeodrilidae, Moniligastridae, Glossoscolecidae, Lumbricidae, Megascolecidae, Eudrilidae.

Sub-class Hirudinea

Body composed of a small prostomium without appendages, and a fixed number (33) of segments which may be secondarily annulated. No chaetae except in Acanthobdellidae. Circum-oral sucker and posterior ventral sucker with the anus opening above it. Clitellum formed on segments 9—11. Testes numerous, in segments posterior to that containing the ovaries. Coelom almost completely obliterated by connective tissue (botryoidal tissue). Carnivorous or ectoparasitic. Mostly in freshwater, but a few marine.

Order 1. Acanthobdellae. Small group in several respects intermediate between Hirudinea and Oligochaeta, in which they are sometimes included. Chaetae retained on segments 2—6, coelom not entirely obliterated, and septa are retained. No anterior sucker. Family: Acanthobdellidae.

Order 2. Rhynchobdellae. Possess an eversible proboscis and lack jaws. Blood vessels retained. Parasites of fishes, amphibians and reptiles. Families: Glossiphoniidae, Piscicolidae.

Order 3. Gnathobdellae. No eversible proboscis, but possess jaws, many species are blood suckers and parasites of birds and mammals. Blood vessels absent and blood circulates in coelomic sinuses. Families: Hirudidae, Haemadipsidae, Semiscolecidae.

Order 4. Pharyngobdellae. No jaws or teeth. Gut without caeca. Exclusively carnivorous. Families: Erpobdellidae, Trematobdellidae, Americobdellidae, Xerobdellidae.

PHYLUM TARDIGRADA

Small animals not exceeding 1 mm in length, with vaguely arthropodan features. Bilaterally symmetrical, with 4 pairs of ventro-lateral legs armed with terminal claws. Cuticle generally thickened and forming a series of plates investing the body and giving the animal a segmented appearance (the cuticle thin and smooth in more primitive tardigrades). 4 pairs of transitory coelomic pouches formed by enterocoely during embryonic development, but do not persist into the adult. Cuticle shed at intervals during growth (ecdysis).

Order 1. Heterotardigrada. A pair of dorso-lateral cirri immediately behind the head. Example: *Echiniscus.*

Order 2. Eutardigrada. No lateral cirri immediately behind the head. Example: *Hypsibius.*

PHYLUM PENTASTOMIDA (= LINGUATULIDA)

Parasitic, chiefly in the respiratory passages of air-breathing vertebrates. Mouth flanked by 2 pair claws on small protuberances, resembling the appendages of tardigrades and onychophorans. Remainder of the body (abdomen) secondarily annulated. Chitinous cuticle not markedly thickened, but shed periodically at ecdysis.

Order 1. Cephalobaenida. Ventral nerve cord with 4 ganglia. Male and female genital openings at anterior end of abdomen. Example: *Cephalobaena*.

Order 2. Porocephalida. Ventral ganglia fused together and with the supra-oesophageal ganglion to form a single anterior ganglion mass. Male gonopore at anterior end of abdomen, female gonopore near posterior end. Example: *Linguatula*.

PHYLUM ONYCHOPHORA

Soft-bodied segmented animals with marked arthropodan features. 14—43 segments each with a pair of ventral legs armed with claws. Head with a pair of antennae. Body cavity a haemocoel. Tracheae. Striated muscles. Terrestrial. Example: *Peripatus*.

PHYLUM ARTHROPODA

Bilaterally symmetrical, metamerically segmented Metazoa, with paired appendages on some or all of the somites. Chitinous cuticle generally thickened and calcified or tanned to form an exoskeleton with thinner, more flexible regions serving as joints between the somites and often on the appendages to form jointed legs. The coelom is reduced and confluent with the blood vascular system, forming a haemocoel. A number of pre-oral segments fused to form a head. The posterior, terminal element, when present, is known as a telson.

The great majority of animals are arthropods. Since so many animals have to be classified within a single phylum, it has proved necessary to introduce a number of additional taxa: sub-phylum, division, super-order, etc. This results in a very extensive and complicated classificatory system, but it is a logical one. Grouping of classes into sub-phyla is common but has not been universally adopted.

SUB-PHYLUM TRILOBITOMORPHA

The fossil trilobites. Body divided into three regions: a solid, anterior cephalon composed of the acron and four fused segments; a middle trunk or thorax composed of numerous segments; and a terminal 'pygidium' composed of a variable number of fused segments. The cephalon covered by a dorsal shield-like carapace. Trilobites had one pair of antennae, and all post-antennal segments bore a pair of biramous appendages.

SUB-PHYLUM CHELICERATA

Body divided into two regions: a prosoma composed of the preoral segments fused with 6 or 7 postoral segments: and an opisthosoma composed of the remaining segments, sometimes reduced. A single pair of preoral appendages (originally postoral) modified as prehensile chelicerae, the second pair of appendages (postoral) generally ambulatory (pedipalps), but may be modified as sensory or prehensile organs.

Class Merostomata
Aquatic, gill-breathing chelicerates with a prosoma bearing the chelicerae and walking legs, and an opisthosoma, 5 or 6 segments of which bear appendages modified as gills. The remaining segments of the opisthosoma lack appendages.
Order 1. Eurypterida. Giant fossil chelicerates. Short, non-segmented prosoma bearing six pairs of appendages, including the chelicerae. Elongated, segmented opisthosoma, the first 5 segments of which bear gills, 7 additional segments without appendages, and a telson.
Order 2. Xiphosura. Body heavily chitinized and with indistinct segmentation. Prosoma covered by a horseshoe-shaped carapace and bearing chelicerae and 5 pairs of walking legs; separated from the opisthosoma by a hinge. Opisthosoma with 7 pairs of appendages, the first rudimentary, the second forming an operculum, the remaining 5 pairs modified as gills. The long caudal spine articulates with the opisthosoma and represents a number of reduced and fused segments and the telson. Example: *Limulus*.

Class Arachnida
Mostly terrestrial and carnivorous, though a few are secondarily aquatic. Prosoma with 6 pairs of appendages: chelicerae, pedipalps (sensory or prehensile) and 4 pairs of walking legs. Opisthosoma primitively of 13 segments and a telson, but is frequently reduced. A variety of respiratory devices formed on anterior opisthosomal segments, but never gills.

Order 1. Scorpionones (scorpions). Prosomal segments fused and with a dorsal carapace. Opisthosoma divided into an anterior mesosoma of 7 segments and a metasoma of 5 segments plus a telson which bears a poison spine. The first opisthosomal segment is missing in scorpions so that the genital openings appear on the first segment of the mesosoma. The second mesosomal segment bears a pair of sensory pectens. Example: *Scorpio*.

Order 2. Pseudoscorpiones (pseudoscorpions). Small arachnids superficially resembling scorpions, but retaining the pre-genital opisthosomal segment. The opisthosoma is not divided into two parts. Example: *Chelifer*.

Order 3. Solpugida (false spiders or sun spiders). Prosoma and opisthosoma segmented and not divided from each other by a pedicel. No telson. Chelicerae very large, pedipalps prehensile, the first pair of walking legs modified as tactile organs and used as antennae. Very extensive tracheal system. Example: *Galeodes*.

Order 4. Palpigradi. Very small arachnids. Opisthosoma joined to prosoma without a pedicel. The terminal segment of the opisthosoma bears a long, articulated flagellum. Chelicerae well developed, pedipalps unmodified and used as walking legs, first pair of walking legs modified as tactile organs and used as antennae. No respiratory organs. Example: *Koenenia*.

Order 5. Uropygi (whip scorpions). Pedipalps prehensile, first walking legs modified as antennae. Opisthosoma composed of 12 segments, the first forming a pedicel by which the opisthosoma is attached to the prosoma, the last three also narrow and constituting a small post-abdomen. The terminal segment bears a jointed flagellum which is often very long.

Sometimes the two sub-orders Holopeltida and Schizopeltida are elevated to the rank of separate orders. The prosoma of the former is covered by a thick chitinous shield in one piece, the prosoma of the latter, which are very small animals, is covered by thin, separate chitinous plates. There are also important internal differences. Examples: *Mastigoproctus, Thelyphonus, Schizomus*.

Order 6. Amblypygi. Pedipalps prehensile, first walking legs modified as antennae and are enormously long and slender. Opisthosoma of 12 segments of which the first forms a pedicel by which the opisthosoma is attached to the prosoma. The terminal segment is narrow but does not bear a flagellum. Examples: *Charinus, Phrynichus*.

Order 7. Araneida (spiders). Prosoma separated from the non-segmented opisthosoma by a pedicel. Chelicerae with poison glands. Pedipalps prehensile and modified as copulatory organs in the male. 4 pairs of walking legs. Opisthosoma soft and bears 2 or 4 pairs of spinning glands. Examples: *Tegenaria, Atypus*.

Order 8. Ricinulei. Cuticle thick with granulations and warts. Prosoma with a hood that can be lowered over the mouth and chelicerae. Third pair of legs modified as copulatory organs in the male. Segments of opisthosoma completely fused and heavily sclerotized. Terminal segment forms a peduncle on which the anus opens. Example: *Ricinoides*.

Order 9. Opiliones (harvest men). No constriction between the prosoma and opisthosoma, which form a single mass. Opisthosoma distinctly segmented. 4 pairs of very long walking legs. Male with a penis, female with an ovipositor. Examples: *Phalangium, Oligolophus*.

Order 10. Acari (mites and ticks). Prosoma and opisthosoma completely fused, with a total loss of segmentation, though secondary segments may sometimes be formed. Chelicerae and pedipalps highly variable in form and function, depending upon the feeding habits of the species. 2, 3 or 4 pairs of walking legs, most commonly 3 pairs in the larva and 4 pairs in the adult. The Acari is a very large, varied and possibly polyphyletic group almost impossible to define categorically. Examples: *Ixodes, Tyroglyphus*.

Class Pycnogonida
Marine chelicerates with a long, slender, segmented body differently organized from that of other chelicerates. An anterior cephalon and proboscis bears 3 pairs of ventral appendages: chelicerae, palps and ovigerous legs (Fig. 9). A trunk composed of 4—6 segments, each bearing a pair of long walking legs. A short, unsegmented abdomen.

Order 1. Colossendeomorpha. Body segments sometimes coalescent. Chelicerae may be rudimentary or even lacking; if present, have a scape (basal section) with 2 joints. Palps with 8—10 joints; ovigerous legs present in both sexes and have 10 joints. 4, 5 or 6 pairs of walking legs. Genital openings on all pairs of legs in both sexes or only on the last two. Examples: *Dodecolopoda, Colossendeis*.

Order 2. Nymphonomorpha. Cephalon generally separated from the trunk by a distinct neck. Chelicerae rarely missing and almost always

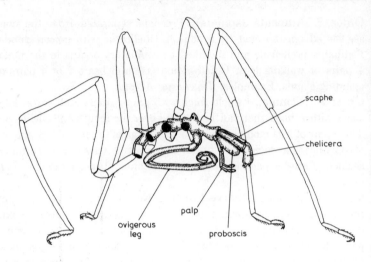

FIGURE 9   Pycnogonid features (after Hedgpeth).

with 1- or 2-jointed scape. Palps with more than 7 joints. Ovigerous legs often rudimentary or missing, especially in the female. Genital openings on all the walking legs in the female, only on the last 2 or 3 pairs in the male. Examples: *Nymphon, Phoxichilidium.*

Order 3. Ascorhynchomorpha. Proboscis very long, at least as long as the trunk. Ovigerous legs in both sexes, but palps and chelicerae variable. Examples: *Ascorhynchus, Achelia.*

Order 4. Pycnogonomorpha. No chelicerae or palps. Females lack ovigerous legs and those of the male lack modified spines but terminate in a hook. Example: *Pycnogonum.*

SUB-PHYLUM MANDIBULATA

Body form variable. First pair of appendages and, if present, the appendages of the succeeding segment, modified as antennae. The next pair of appendages modified as jaws (mandibles) and one or two pairs of succeeding appendages modified as additional jaws (maxillae).

Class Crustacea

Arthropods with 2 pairs of antennae and 3 pairs of mouthparts (mandibles, 1st maxillae, 2nd maxillae). Usually with a carapace. Number

of segments generally fixed. The majority of crustaceans are aquatic
and gill-breathing.

Sub-class Cephalocarida
Very small crustaceans 3—4 mm long, with small horseshoe-shaped
cephalic carapace and 19—20 post-cephalic segments, including the
telson. All cephalic appendages present, the second maxillae similar
to the thoracic limbs with which they are serially arranged. 7 pairs of
thoracic appendages and an eighth pair reduced or absent. Abdominal
segments without limbs. Basal section of trunk appendages (Fig. 10)
and second maxilla with large, flattened outer 'pseudepipodite' giving
the limb a triramous structure. (The Cephalocarida are sometimes
regarded as an order of branchiopods.) Examples: *Hutchinsoniella*,
Lightiella (only known genera).

Sub-class Branchiopoda
Carapace forming a dorsal shield or bivalved shell, or may be absent.
Trunk appendages usually foliaceous (Fig. 10), at least 4 pairs and may
be numerous. Posterior part of body without appendages and ends in
a caudal furca. Mandibular palp reduced or absent. Maxillae absent or
reduced. Paired eyes usually present. Young hatch as nauplii or me-
tanauplii.
Order 1. Anostraca. No carapace. Paired eyes pendunculate. Trunk

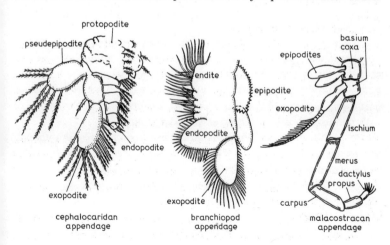

FIGURE 10   Types of crustacean thoracic appendages.

limbs 11, 17 or 9, more post-genital. 8 or 9 segments without limbs. Unsegmented caudal furca. Development with metamorphosis. Examples: *Artemia, Chirocephalus, Branchipus*.

Order 2. Lipostraca. An extinct order with one species known from Middle Devonian. Resembled Anostraca but had no eyes. Antennae biramous. First maxillae in male modified as claspers. Trunk limbs in 2 series, first 3 for feeding, remainder for swimming. Example: *Lepidocaris*.

Order 3. Notostraca. Carapace forms a dorsal shield. Paired eyes sessile, and coalescent antennae vestigial. More than 40 pairs of trunk limbs, 29—52 being post-genital. Development with metamorphosis. Examples: *Triops* (= *Apus*), *Lepidurus*.

Order 4. Conchostraca. Carapace forms a bivalved shell which encloses head and body. Paired eyes sessile, coalescent. Antennae biramous, used for swimming. 10—32 pairs of trunk limbs 0—16 being post-genital. Rami of caudal furca claw-like. Development usually with metamorphosis. Example: *Estheria*.

Order 5. Cladocera. Carapace forms a bivalved shell which encloses the body but not the head; it may be reduced and serve only as a brood-pouch. Paired eyes sessile and coalesced. Antennae biramous, used for swimming. 4—6 pairs of trunk limbs, none post-genital. Rami of caudal furca claw-like. Development usually without metamorphosis. Examples: *Daphnia, Simocephalus, Podon, Evadne*.

Sub-class Ostracoda

Carapace forms a bivalved shell which encloses head and body. Paired eyes sometimes present, sessile. First and second antennae large and used for locomotion. Mandibles with palp. Not more than 4 pairs of limbs behind mandibles. Trunk indistinctly segmented, the posterior part being without limbs and ending in a caudal furca. Development with metamorphosis, hatching as a modified nauplius.

Order 1. Myodocopa. Shell has an antennal notch. Second antennae usually biramous and have a very large basal segment. Two pairs of trunk appendages. Examples: *Cypridina, Conchoecia*.

Order 2. Cladocopa. Second antenna biramous, no trunk appendages. Marine ostracods. Example: *Polycope*.

Order 3. Podocopa. Second antenna uniramous. Two pairs of trunk appendages. Marine and freshwater ostracods. Examples: *Cypris, Candona, Darwinula, Cythere*.

Order 4. Platycopa. Second antenna biramous. One pair of trunk appendages. Contains a single genus of marine ostracods. Example: *Cytherella.*

Sub-class Copepoda

No carapace. A median 'nauplius' eye in the adult, but no paired eyes. Body broad anteriorly, tapering posteriorly and distinctly segmented, terminating in a caudal furca. Typically one segment fused with the head and 9, including the telson, are free, although some degree of fusion between these may occur and an additional segment may be fused with the head. First antennae large, uniramous and used for swimming. 4 pairs of mouthparts: mandibles, first and second maxillae, maxillipeds (1st thoracic limbs). 4 or 5 pairs of trunk limbs in addition to maxillipeds. 4 abdominal segments without appendages. Articulation between the fourth and fifth or between the fifth and sixth post-cephalic segments is much greater than that between the other somites and is used to divide the copepod body into a prosoma (head and thoracic segments fused with it), metasoma (free thoracic segments up to articulation) and urosoma (remainder of thorax and abdominal segments). Nauplius stage present. (Most of these characters are modified or missing in the parasitic forms.)

Order 1. Calanoida. Metasome-urosome articulation between fifth and sixth post-cephalic segments. Second antennae biramous. Free-living, generally planktonic copepods. Examples: *Calanus, Diaptomus, Metridia, Euchaeta, Centropages.*

Order 2. Harpacticoida. Metasome-urosome articulation between fourth and fifth post-cephalic segments. First antennae very short and second antennae biramous. Abdomen almost as wide as thorax. Generally benthic, in fresh- or sea-water; a few parasitic. Examples: *Harpacticus, Tigriopus, Euterpina, Metis.*

Order 3. Cyclopoida. Metasome-urosome articulation between fourth and fifth post-cephalic segments. Second antennae uniramous. Includes planktonic and benthic forms living in fresh- or sea-water; many are parasitic. Examples: *Cyclops, Sapphirina, Oncaea, Lernaea.*

Order 4. Notodelphyoida. Metasome-urosome articulation between fourth and fifth post-cephalic segments in males and between first and second abdominal segments in females. Live as commensals in tunicates: not highly modified. Examples: *Ascidicola, Notodelphys, Doropygus.*

Order 5. Monstrilloida. Second antennae and mouthparts lacking in adult. Includes marine copepods with larval stages parasitic in poly-chaetes. Examples: *Monstrilla, Xenocoeloma.*

Order 6. Caligoida. Metasome-urosome articulation between third and fourth post-cephalic segments, but this articulation may be absent in the more highly modified females. Second antennae modified for attachment to the host. Mostly ectoparasitic on freshwater and marine fish. Examples: *Caligus, Lepeophtheirus, Peroderma, Penella Lerna-eocera* (= the parasitic *'Lernaea'* of textbooks).

Order 7. Lernaeopodoida. Segmentation reduced or absent. Thoracic appendages reduced or absent, especially in the female. Second maxil-lae modified for attachment to the host. Ectoparasitic on freshwater and marine fish. Examples: *Salmincola, Penella, Brachiella, Lesteira.*

Sub-class Mystacocarida

Elongated cylindrical body divided as in Copepoda, except that first thoracic segment is not fused with the head. Thorax of 4 segments each with a pair of appendages consisting of a single simple lamella.

Order 1. Derocheilocarida. Example: *Derocheilocaris* (only known genus).

Sub-class Branchiura

Parasites of freshwater and marine fish. Compound eyes present as well as median eye. Body depressed and divided into unsegmented cephalo-thorax, 3 free thoracic segments and an unsegmented abdominal region. Thoracic limbs 2—5 biramous and used for swimming. A hollow, protusible 'poison spine' in front of mouth. Development direct. Examples: *Argulus, Dolops.*

Sub-class Cirripedia

Sessile as adults. Carapace present (except in *Proteolepas*) as a mantle enclosing whole of body and limbs; usually strengthened by calcareous plates. Paired eyes absent in adult. First antennae organs of attachment which become vestigial in adult. 6 pairs of biramous, cirriform limbs used for feeding. Hatch as nauplius. Some of these characters do not apply to parasitic species.

Order 1. Thoracica. Mantle present. 6 pairs of cirriform trunk ap-pendages. Free-living and mantle generally covered with calcareous plates. Examples: *Lepas, Pollicipes, Scalpellum, Conchoderma, Balanus, Chthamalus, Verruca.*

Order 2. Acrothoracica. Mantle present. Trunk appendages reduced in number, the posterior pairs separated from the first. Examples: *Alcippe, Trypetesa.*

Order 3. Ascothoracica. Mantle present and containing diverticula of gut. Trunk limbs reduced but prehensile, first antenna and abdomen usually present. Parasitic. Examples: *Laura, Dendrogaster, Ascothorax.*

Order 4. Apoda. Mantle present. No trunk limbs. Parasitic. Example: *Proteolepas* (only known genus).

Order 5. Rhizocephala. Mantle present. Appendages and alimentary canal absent. Parasitic, primarily in decapod crustaceans, forming absorptive 'roots'. Examples: *Sacculina, Peltogaster.*

Sub-class Malacostraca

Carapace developed to varying extent but never enclosing whole of abdomen and may be completely absent. Usually 14, rarely 15, trunk segments, all (except 15) bearing appendages, and a telson, the latter usually without a caudal furca. Trunk limbs (Fig. 10) in two tagmata, a thoracic of 8 and an abdominal of 6 pairs. Female apertures on 6th, male on 8th trunk segment. Paired eyes usually present. Development usually with metamorphosis but a nauplius rarely present.

SERIES LEPTOSTRACA

SUPER-ORDER PHYLLOCARIDA

Order 1. Nebaliacea. Abdomen of 7 segments, the last without appendages. Caudal furca present. Bivalved carapace with transverse

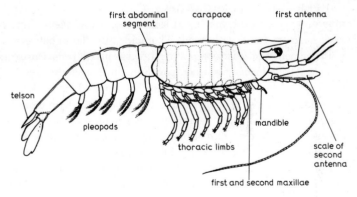

FIGURE 11 General features of a malacostracan ('the caridoid fascies').

adductor muscle. Eyes pedunculate. Thoracic limbs all similar, more or less foliaceous, with 3-segmented protopodite. Example: *Nebalia.*

SERIES EUMALACOSTRACA
Abdomen of 6 segments, all bearing appendages. No caudal furca, carapace without adductor muscle. Eyes pendunculate or sessile. Thoracic limbs rarely all similar; not foliaceous. Protopodite 2-segmented except in Hoplocarida.

SUPER-ORDER SYNCARIDA
Order 1. Anaspidacea. No carapace. 1st thoracic segment fused with head but may be defined by a groove. Thoracic limbs similar in general structure; flexed between 5th and 6th segments; with exopodite, except last 1 or 2 pairs; with lamellar epipodites. No oostegites. Heart elongated. Examples: *Anaspides, Koonunga.*
Order 2. Bathynellacea. Differ from Anaspidacea in that first thoracic segment is free instead of being fused with the head. Examples: *Bathynella, Thermobathynella.*

SUPER-ORDER PANCARIDA
Order 1. Thermosbaenacea. Body nearly cylindrical without a constriction between thorax and abdomen. Carapace fused with 1st segment but covering segments 1—4. Telson fused with 6th abdominal segment. 1 pair of maxillipeds, 5 pairs of biramous thoracic legs, 2 pairs of rudimentary, uniramous abdominal appendages. Uropods biramous. Example: *Thermosbaena.*

SUPER-ORDER PERACARIDA
Carapace, when present, leaving at least 4 thoracic segments free. 1st thoracic segment fused with head. Thoracic limbs flexed between 5th and 6th segments; at least one pair modified as maxillipeds. Oostegites present in female forming a brood pouch in which eggs develop.
Order 1. Mysidacea. Peracarida having the caridoid facies (Fig. 11). Carapace covers most of thorax but not fused dorsally with more than 3 thoracic segments. Eyes pedunculate when present. 1st antennae biramous. 2nd antennae with scale-like exopodites. 1 or 2 maxillipeds. Thoracic limbs with exopodites used for swimming. Statocyst on endopodite of uropod. 2—7 oostegites. Examples: *Praunus, Mysis.*
Order 2. Cumacea. Peracarida in which carapace fused dorsally with

3 or 4 thoracic segments and forms a branchial cavity on each side. Eyes fused and immovable. 2nd antennae without exopodite. 3 pairs of maxillipeds. Pleopods absent in female and often reduced in male. 4 oostegites. Examples: *Pseudocuma, Diastylis.*

Order 3. Tanaidacea. Paracarida in which the body is usually depressed. Carapace fused dorsally with 2 thoracic segments and forms a branchial cavity in each side. Eyes pedunculate but immovable. 2nd antennae may have exopodite. 1 pair of maxillipeds. Pleopods usually present. 1—6 oostegites. Examples: *Apseudes, Tanais.*

Order 4. Gnathiidea. Body depressed. No carapace. First and seventh thoracic segments reduced and second thoracic segment fused with head. Eighth thoracic appendages absent. Male with enormously developed mandibles. Sometimes regarded as a sub-order of isopods. Example: *Gnathia.*

Order 5. Isopoda. Peracarida in which body is typically depressed. No carapace. 1st thoracic segment and rarely also 2nd, fused with head. Telson usually fused with last segment. Eyes sessile. 1st antennae nearly always uniramous. 2nd antennae sometimes with minute exopodite. 1 pair maxillipeds. No exopodites or thoracic limbs. Marine, freshwater and land forms. Many marine forms parasitic. Usually 5 oostegites. Examples: *Asellus, Limnoria, Sphaeroma, Idotea, Ligia, Oniscus, Bopyrus.*

Order 6. Spelaeogriphacea. Body elongated, more or less cylindrical but slightly depressed. Carapace fused with first thoracic segment and overlapping most of 2nd. Thoracic segments 2—8 free. Long abdomen, more than half the total length of the body, with telson free from last segment. Blind, though with movable, oval, plate-like optic lobes. 5 oostegites. Cave-dwelling form from S. Africa. Example: *Spelaeogriphus* (only genus).

Order 7. Amphipoda. Peracarida in which body is typically compressed. No carapace. 1st thoracic segment, and rarely also 2nd fused with head. Telson usually distinct from last segment. Eyes sessile. 1st antennae often biramous. 2nd antennae without exopodite. 1 pair of maxillipeds. No exopodites on thoracic limbs. Usually 4 oostegites. Examples: *Gammarus, Orchestia, Caprella.*

SUPER-ORDER EUCARIDA

Carapace fused dorsally with all thoracic segments. Eyes pedunculate. Thoracic limbs flexed between 4th and 5th segments. No oostegites.

Heart short, thoracic. Development generally without metamorphosis, but a nauplius present in more primitive forms.

Order 1. Euphausiacea. Eucarida with the caridoid facies (Fig. 11). No maxillipeds, all thoracic limbs being similar in structure. Single series of gills (podobranchs) attached to coxopodites of thoracic limbs, not covered by carapace. Young hatched as nauplii. No oostegites. Eggs carried between thoracic legs and sometimes cemented together. Examples: *Euphausia, Nyctiphanes.*

Order 2. Decapoda. Eucarida which often retain caridoid facies but often are greatly modified. 2nd maxilla bears a scaphognathite on outer side. 3 pairs of maxillipeds. Generally several series of gills attached to coxopodites, arthrodial membranes, and body wall of thoracic segments. Young rarely hatched in nauplius stage.

Sub-order 1. Natantia. Body nearly always laterally compressed and generally retains caridoid facies. Rostrum present. Exopodite of 2nd antenna formed into large, flat 'scale'. Full number of well developed pleopods present and used for swimming. Examples: *Pandalus, Hippolyte, Leander, Crangon.*

Sub-order 2. Reptantia. Body not compressed, often depressed. Exopodite of 2nd antenna small or absent. Pleopods often reduced or absent, usually not used for swimming. Examples: *Palinurus, Nephrops, Homarus, Astacus, Galathea, Porcellana, Eupagurus, Lithodes, Carcinus, Cancer, Hyas.*

SUPER-ORDER HOPLOCARIDA

Order 1. Stomatopoda. At least 4 thoracic segments free from carapace. Two movable segments formed from front of head and bearing pedunculate eyes and 1st antennae respectively. Heart extends through thorax and abdomen. Development resembles Decapoda, the egg hatching as a zoaea. Examples: *Squilla, Gonodactylus.*

Class Diplopoda

One pair of antennae. Numerous body segments which, apart from the first 4, are fused together in pairs so that they bear apparently 2 pairs of walking legs each. Genital openings on the 3rd segment behind the head.

Sub-class Pselaphognatha

Small millipedes with a soft, uncalcified integument. Characteristic barbed bristles (trichomes) as well as other bristles, hooks and hairs on the body, giving the animal a generally hairy appearance. Trichobothria on the head. No appendages modified as copulatory organs. Order 1. Polyxenida (the only order in the sub-class, containing a single family, Polyxenidae). Examples: *Polyxena, Lophoproctus.*

Sub-class Chilognatha

Some calcification of integument as well as phenolic tanning of the cuticle. Hairs, when present on the body are smooth; no trichobothria. Male with copulatory organs, either walking legs of the 7th annulus modified as gonopods, or special clasping telopods developed at the posterior end of the body.

SUPER-ORDER 1 PENTAZONIA

Gonopods never developed on the 7th annulus, but 1 or 2 pairs of telopods at the posterior end of the body, the last pair of these being sperm transfer organs.

Order 1. Glomerida (= Oniscomorpha). Short, broad body of 14—17 annuli, including the anal segment, bearing 17—21 pairs of appendages (plus 2 pairs of telopods in the male), with 11—13 tergal plates, the second of which is much larger than the rest. Body can be rolled into a ball concealing the head under the enlarged 2nd tergite. Examples: *Glomeris, Sphaerotherium.*

Order 2. Glomeridesmida (= Limacomorpha). Small millipedes. Body composed of 22 annuli bearing 36 pairs of appendages (plus 1 pair of telopods in the male). Second tergite not enlarged and body cannot be rolled into a ball. Example: *Glomeridesmus.*

SUPER-ORDER 2 HELMINTHOMORPHA

Long-bodied millipedes, spirally coiled when at rest. One or often both pairs of legs on the 7th annulus modified as gonopods. No telopods.

Order 1. Chordeumida. Body cylindrical or with lateral flanges. 26 or more annuli with the first or both pairs of legs of 7th annulus modified as gonopods. No repugnatorial glands. 2—3 pairs of silk glands in the posterior part of the body, opening by 1—3 pairs of spinnerets on the anal segment. Examples: *Chordeuma, Callipus, Stemmiulus.*

Order 2. Polydesmida. Body cylindrical or, more commonly, with lateral flanges. 19—22 annuli with only the first pair of legs of the 7th annulus modified as gonopods, the second pair normal walking legs. With or without repugnatorial glands. No silk glands. No eyes. Examples: *Polydesmus, Oxidus, Orthomorpha, Platyrrhachus.*

Order 3. Juliformia. Body cylindrical with at least 40 annuli. Both, or only the first pair of legs of the 7th annulus modified as gonopods; in the latter case the second pair of legs missing. Repugnatorial glands on all tergal plates except a few at the anterior end. No silk glands.

This is a large order and two of its constituent sub-orders Julida and Spirobolida, and the two divisions, Spirostreptida and Cambalida of the third sub-order, Spirostreptomorpha, are now sometimes all raised to the rank of separate orders. Examples: *Julus, Spirobolus, Trigonolus, Spirostreptus, Cambulus.*

SUPER-ORDER 3 COLOBOGNATHA

The first pair of appendages of the 7th annulus normal walking legs; the second pair of 7th annulus and first pair of 8th annulus modified as gonopods, but the modification is slight. Mouthparts highly modified with toothless labrum and no jaws on the other mouthparts, tending to be reduced and in extreme cases forming a long, piercing, sucking beak. Examples: *Siphoniulus, Platydesmus, Dolistenus.*

Class Chilopoda

One pair of antennae. One pair of walking legs on each of the body segments posterior to the first which bears a pair of poison claws. Genital opening at the posterior end of the body.

Sub-class Epimorpha

Eggs laid in a mass and brooded by the female. Young hatch with full complement of legs.

Order 1. Geophilomorpha. Long, worm-like centipedes with 31 to over 170 pairs of legs. Anterior part of each segment marked off from the posterior part by a distinct joint. Antennae of 14 joints. Spiracles on all segments except the first and last. Burrowing forms. Examples: *Geophilus, Mecistocephalus, Orya, Himantaria.*

Order 2. Scolopendromorpha. Body relatively short and stout, with not more than 23 pairs of legs. No marked division of the somites. Antennae of 17—30 joints. Spiracles not on every segment. Examples: *Scolopendra, Cryptops.*

Sub-class Anamorpha

Eggs laid singly and not brooded. Young hatch with 7 pairs of legs and the additional legs added in about 4 'larval' moults.

Order 1. Lithobiomorpha. 15 pairs of legs. Antennae with 20—50 segments. 6—7 pairs of spiracles; terga of those segments without spiracles reduced in size. Examples: *Lithobius, Etholpolys, Henicops, Craterostigmus.*

Order 2. Scutigeromorpha. Long-legged, agile, fast-running centipedes, with bulging compound eyes. 15 pairs of legs and sternal plates, but only 8 tergal plates. Spiracles unpaired, opening dorsally on the first 7 tergites. Example: *Scutigera.*

Class Symphyla

One pair of antennae, 14 body segments, with 12 walking legs and a 13th preanal segment with cerci. Genital openings on the 3rd segment behind the head. Example: *Scutigerella.*

Class Pauropoda

One pair of antennae, one pair of maxillae, 12 body segments, of which the 1st, 11 and 12th are without appendages. Others each with a pair of walking legs. Genital opening on the 3rd segment behind the head. Example: *Pauropus.*

Class Insecta

One pair of antennae. Body divided into head, thorax and abdomen. Thorax of three segments bearing walking legs (Fig. 12) and usually 1 or 2 pairs of wings (Fig. 13). Abdomen without walking legs. Genital opening near posterior end of body.

Sub-class Apterygota

Wingless insects (the wingless condition presumed to be primitive). Metamorphosis slight or absent. Adult with one or more pairs of pre-

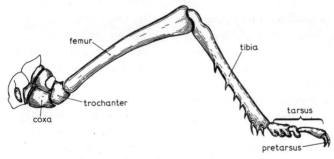

FIGURE 12   Insect leg (grasshopper).

genital abdominal appendages. Adult mandibles usually articulating with the head capsule at a single point.

Order 1. Thysanura (bristle tails, silver fish). With biting, ectognathous mouthparts. Antennae with many segments but only the basal segment provided with muscles. Tarsi with 2—4 segments. Abdomen of 11 segments, with a variable number of lateral styliform appendages. Terminal segment with a pair of cerci composed of many segments and a median, terminal process. Tracheal system and Malpighian tubules. Examples: *Lepisma, Petrobius, Machilis, Thermobia.*

Order 2. Diplura. With biting entognathous mouthparts. Antennae with many segments, all provided with muscles. Tarsi with 1 segment. Abdomen with 11 segments, the last very small, with lateral styliform appendages on most or all of the pregenital segments. Terminal segment with a pair of cerci of variable form, but without a median process. Tracheal system, but Malpighian tubules vestigial or absent. Examples: *Campodea, Japyx.*

Order 3. Protura. Minute insects (less than 2 mm long) with entognathous mouthparts. No antennae or compound eyes. Abdomen with 11 segments, the last with a well-developed telson. First three segments with a pair of small appendages. Malpighian tubules reduced to papillae. Metamorphosis slight (chiefly addition of abdominal segments). Examples: *Acerentomon, Eosentomon.*

Order 4. Collembola (spring tails). With biting entognathous mouthparts. Antennae of 4 segments, the first 3 with muscles. No compound eyes. Abdomen of 6 segments with 3 pairs of appendages: a ventral tube on segment I, a small retinaculum on III and a forked springing organ on IV. Tracheal system usually absent. No Malpighian tubes. No metamorphosis. Examples: *Anurida, Protura, Sminthurus, Neelus.*

Sub-class Pterygota

Winged insects (if wingless this is secondary). Metamorphosis varies but rarely slight or lacking. Adult without pregenital abdominal appendages. Adult mandibles usually articulating with the head capsule at two points.

The classification given below is conservative and is the one used by Imms *A General Textbook of Entomology* (9th edit, revised by O.W. Richards and R. G. Davies, Methuen, London 1957) and most other textbooks. An alternative grouping of the orders into divisions and sections is also in use (Jeannel *Traité de Zoologie* (ed. P. P. Grassé)

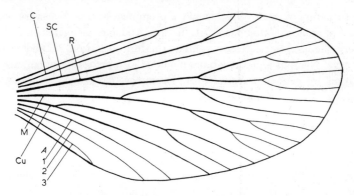

FIGURE 13 Generalized venation of an insect wing (after Imms).

9, 3—17. Masson, Paris, 1949), and involves the replacement of the
Exopterygota and Endopterygota by two new divisions, the Palaeoptera and Neoptera, the latter in three sections: Polyneoptera,
Paraneoptera and Oligoneoptera. The Oligoneoptera correspond
exactly with the Endopterygota. The exopterygote orders are divided
between the Palaeoptera and the first two sections of the Neoptera.
These alternative groupings of the orders are indicated in square
brackets.

DIVISION 1 EXOPTERYGOTA (= HEMIMETABOLA)
Metamorphosis simple and sometimes slight. Pupal instar rarely present. Wings develop externally. Immature stages generally nymphs
which resemble adults in structure and habit.

[ALT. DIVISION 1. PALAEOPTERA]
Order 1. Ephemeroptera (mayflies). Soft-bodied insects with short,
bristled antennae and vestigial mouthparts. Wings membranous, held
vertically upwards when at rest, the hind pair considerably reduced.
Intercalary veins and numerous cross veins. Abdomen terminates in
a pair of very long cerci sometimes with a similar median process as
well. Nymphs aquatic usually with 3 long caudal appendages, and gills.
Adult preceded by a sub-imaginal winged instar. Examples: *Ephemera,
Baetis, Ecdyonurus*.
Order 2. Odonata (dragonflies). Predaceous insects with biting
mouthparts. 2 pairs equal elongated membranous wings; with complex

reticulation of small veins; usually a prominent stigma. Eyes very large and prominent. Antennae short, filiform. Abdomen elongated and often very slender. Male accessory genital armature on 2nd and 3rd abdominal sterna. Nymphs aquatic; labium modified as a prehensile organ; respiration by caudal or rectal gills. Examples: *Cordulegaster, Aeshna, Agrion, Lestes, Epiophlebia.*

[ALT. DIVISION 2. NEOPTERA

SECTION 1. POLYNEOPTERA]

Order 3. Plecoptera (stoneflies). Soft-bodied insects with elongate bristled antennae. Weak biting mouthparts. Mandibles normal or vestigial, ligula 4-lobed (Fig. 14). Wings membranous held flat over back when at rest; hind pair generally the larger, with well developed anal lobes. Venation variable and often specialized, vein M 2-branched. Tarsi 3-segmented. Abdomen usually terminates in long, multiarticulated cerci. No ovipositor. Nymphs aquatic; with antennae and usually cerci elongate; with tracheal gills as a rule, position variable. Examples: *Perla, Nemoura, Nephelopteryx.*

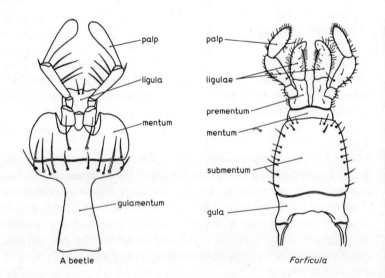

FIGURE 14   Insect labrum (ental view) with single or divided ligula (after Imms).

Order 4. Grylloblattodea. Without wings. Eyes reduced or absent, no ocelli. Antennae moderately long and filiform. Mouthparts mandibulate. Legs similar to each other, tarsi with 5 segments. Female with well developed ovipositor. Male genitalia asymmetrical, cerci long, with 8 segments. Example: *Grylloblatta* (only 6 species in the order).

Order 5. Orthoptera (grasshoppers, locusts, crickets). Wings variable and sometimes absent. Mouthparts mandibulate. Prothorax large. Hind legs usually enlarged and modified for jumping. Coxae small and rather widely separated. Tarsi usually with 3 or 4 segments. Forewings generally thickened with submarginal costal vein. Wing pads of nymph undergo reversal during development. Female generally with well-developed ovipositor not concealed by 7th or 8th abdominal sterna. Male external genitalia symmetrical, concealed at rest by enlarged 9th abdominal sternum which sometimes bears a pair of styles. Cerci usually short and nearly always segmented. Specialized auditory and stridulatory organs as a rule. Metamorphosis slight. Examples: *Locusta, Schistocerca, Tettigonia, Gryllotalpa, Achaeta, Oecanthus*.

Order 6. Phasmida (stick insects, leaf insects). With or without wings. Either flattened and leaf-like or, more often, elongated and cylindrical. Mouthparts mandibulate. Prothorax short: meso- and meta-thorax usually long. Legs similar to each other; coxae small and rather widely separated. Tarsi usually with 5 segments. Forewings, if present, usually small and with sub-marginal costal vein. Wing pads of nymphs do not undergo reversal during development. Ovipositor small and mostly concealed by enlarged 8th abdominal sternum. Male external genitalia variable and asymmetrical, concealed by 9th abdominal segment. Cerci short and unsegmented. No specialized auditory and stridulatory organs. Metamorphosis slight. Examples: *Carausius, Phyllium*.

Order 7. Dermaptera (earwigs). Elongated insects with biting mouth-parts; superlinguae present; ligula 2-lobed (Fig. 14). Forewings short, leathery and without veins. Hind-wings semicircular, membranous, with veins largely radial. Wingless forms common. Tarsi with 3 segments. Cerci unjointed and almost always modified into heavily sclerotized forceps. Ovipositor reduced or absent. Metamorphosis slight or wanting. Examples: *Forficula, Arixenia, Hemimerus*.

Order 8. Embioptera. Gregarious insects living in silken tunnels. Biting mouthparts; ligula 4-lobed. Tarsi with 3 segments; 1st segment of anterior pair greatly inflated. Females without wings. Males with both pairs of wings alike, veins little pronounced; R greatly thickened, other veins reduced or absent (Fig. 13). Cerci with 2 segments, generally asymmetrical in male. Metamorphosis gradual in male, absent in female. Examples: *Embia, Oligotoma, Dictyoploca*.

Order 9. Dictyoptera (cockroaches, mantids). Antennae almost always filiform with numerous segments. Mouthparts mandibulate. Legs similar to one-another, or forelegs modified for grasping; coxae large and rather close together, tarsi with 5 segments. Forewings thickened and with marginal costal vein. Wing pads of nymph do not undergo reversal during development. Female with reduced ovipositor concealed by enlarged 7th abdominal sternum. Male genitalia complex, asymmetrical and largely concealed by 9th abdominal sternum which bears a pair of styles. Cerci auditory or stridulatory organs. Eggs concealed in ootheca. Examples: *Blatta, Periplaneta, Blabarus, Blatella, Mantis, Empusa*.

Order 10. Isoptera (termites). Social and polymorphic species living in large communities composed of reproductive forms and numerous wingless, sterile soldiers and workers. Biting mouthparts, ligula 4-lobed. Wings very similar to each other, elongated and membranous, superposed flat on back when at rest, and capable of being shed by basal fractures; anterior veins strongly sclerotized, without regular cross veins, though often with an irregular network of veins (= an archedictyon) between principal ones. Tarsi nearly always with 4 segments. Genitalia generally rudimentary or absent in both sexes.. Metamorphosis slight or absent. Examples: *Archotermopsis, Odontotermes, Kalotermes*.

Order 11. Zoraptera. Winged or wingless. Antennae of 9 segments, the segments like beads on a string. Labial palps with 3 lobes. Wings, if present, capable of being shed by basal fractures; venation specialized by reduction. Prothorax well developed. Tarsi with 2 segments. Cerci very short, with 1 segment. No ovipositor. Male genitalia specialized, sometimes asymmetrical. Metamorphosis slight. Example: *Zorotypus* (Small order containing only 16 species).

[ALT. SECTION 2. PARANEOPTERA]

Order 12. Psocoptera (book lice). Small insects with long filiform antennae with 15—20 segments. Y-shaped epicranial suture, postclypeus enlarged. Maxilla with rod-like lacinia; labial palps reduced, with 1—2 segments. Prothorax generally small. Tarsi with 2 or 3 segments. No cerci. Examples: *Peripsocus, Psocus*.

Order 13. Mallophaga (biting lice, bird lice). Wingless, living as ectoparasites of birds or, less frequently, mammals. Eyes reduced, no ocelli. Antennae with 3—5 segments. Mouthparts of modified biting type; maxillary palps with 4 segments or absent; ligula undivided or with 2 lobes; labial palps rudimentary. Prothorax obvious, meso- and meta-thorax often imperfectly separated. Tarsi with 1 or 2 segments, terminating in single or paired claws. Thoracic spiracles ventral. No cerci. No metamorphosis. Sometimes included with Anoplura as suborders of the order Phthiraptera. Examples: *Lipeurus, Menapon.*

Order 14. Anoplura (= Siphunculata, sucking lice). Wingless, living as ectoparasites of mammals. Eyes reduced or absent. No ocelli. Antennae with 3—5 segments. Mouthparts highly modified for piercing and sucking, retracted into head when not in use. Thoracic segments fused. Tarsi with 1 segment and single claw. Thoracic spiracles dorsal. No cerci. No metamorphosis. Sometimes included with Mallophaga as sub-orders of the order Phthiraptera. Examples: *Pediculus, Phthirus.*

Order 15. Hemiptera (plant bugs). Usually 2 pairs of wings, anterior usually harder than posterior pairs, uniformly so (Homoptera) or with the tip more membranous than the rest (Heteroptera). Mouthparts piercing and suctorial, palps atrophied; labium in the form of a dorsally grooved sheath receiving the two pairs of bristle-like stylets (modified mandibles and maxillae). Metamorphosis usually gradual, rarely complete. Sometimes the two sub-orders are elevated to the rank of separate orders. Examples: Homoptera: *Cicada, Aphis*; Heteroptera: *Cimex* (bed bug), *Corixa* (water boatman), *Dysderca.*

Order 16. Thysanoptera (thrips). Small, slender-bodied insects with antennae of 6—10 segments. Mouthparts asymmetrical, piercing; with maxillary and labial palps. Prothorax well marked, free. Tarsi with 1—2 segments, each with terminal protrusible vesicle. Wings, when present, very narrow with reduced venation and long marginal setae. No cerci. Metamorphosis accompanied by 1 or 2 inactive pupal instars. Example: *Thrips.*

DIVISION 2. ENDOPTERYGOTA (= HOLOMETABOLA)
Metamorphosis is complex and accompanied by a pupal instar. Wings develop internally. Immature stages are larvae which differ from adult in structure and habits.

[ALT. SECTION 3. OLIGONEOPTERA]

Order 1. Neuroptera (alder flies, lacewings, ant lions). Soft-bodied and with elongated antennae as a rule. Mouthparts adapted for biting; ligula undivided or bilobed or often atrophied. Two pairs of very similar membranous wings, generally held like a peaked roof over abdomen when at rest. Venation primitive but with many accessory veins; costal veinlets numerous; R's often pectinately branched. No cerci. Larvae carnivorous with biting or sucking mouthparts; aquatic forms generally with abdominal gills. Pupae usually exarate. Wings with complete tracheation. Examples: *Sialis, Raphidia, Chrysopa, Myrmeleon.*

Order 2. Mecoptera (scorpion flies). Slender, usually carnivorous, with long, filiform antennae. Head usually produced into a ventrally deflected rostrum, with biting mouthparts. No ligula. Legs long and slender. Wings similar and membranous, carried longitudinally and horizontally when at rest; venation primitive, R's dichotomously branched, Cu_1 simple. Abdomen elongated, with short cerci, male genitalia prominent. Larvae with biting mouthparts and 3 pairs of legs; with or without abdominal feet. Pupae exarate; wings with reduced tracheation. Examples: *Panorpa, Bittacus, Boveus.*

Order 3. Lepidoptera (butterflies and moths). With 2 pairs of membranous wings, few cross veins. Body, wings and appendages covered with broad scales. Mandibles nearly always reduced or absent, principal mouthparts consist of a long, suctorial proboscis formed by maxillae. Larvae with 8 pairs of legs as a rule. Pupae more or less obtect and generally enclosed in a cocoon or earthen cell; wing tracheation complete. Examples: *Pieris, Bombyx, Tortrix, Ephesia, Papilio.*

Order 3a. Zeugloptera. This sub-order of Lepidoptera, which unlike the rest has functional mandibles, is sometimes regarded as constituting a separate order. Example: *Micropteryx.*

Order 4. Trichoptera (caddis flies). Moth-like insects with bristled antennae. Mandibles vestigial or absent; maxillae with 1 lobe and elongated palps; labium with median glossa and well developed palps. Wings membranous, more or less densely hairy, held like peaked roof over abdomen when at rest; forewings elongated, hind wings broader and with folding anal area; venation generalized with few cross-veins. Tarsi with 5 segments. Larvae aquatic and live in constructed cases; body terminates in hooked, caudal appendages. Pupae with strong mandibles; wing venation reduced. Examples: *Hydropsyche, Limnephilus, Macronema.*

Order 5. Diptera (true flies). With a single pair of membranous wings, hind wings modified to form halteres. Mouthparts suctorial usually forming a proboscis and sometimes adapted for piercing; mandibles usually absent; labium generally distally expanded in a pair of fleshy lobes. Prothorax and metathorax small and fused with large mesothorax. Tarsi commonly with 5 segments. Metamorphosis complete. Larvae without legs, frequently with head reduced and retracted, tracheal system variable. Pupa either free or enclosed in hardened larval cuticle, wing tracheation reduced. Examples: *Anopheles* (mosquito), *Musca* (house fly), *Simulium, Chironomus, Tipula, Drosophila, Calliphora*.

Order 6. Siphonaptera (fleas). Adult wingless and laterally compressed, ectoparasites of warm-blooded vertebrates. Eyes absent, but usually with 2 ocelli. Antennae short and stout, reposing in grooves. Mouthparts modified for piercing and sucking; with maxillary and labial palps. Thoracic segments free with very large coxae. Tarsi with 5 segments. Larvae elongated and without legs. Pupae enclosed in cocoons. Examples: *Pulex, Tunga*.

Order 7. Hymenoptera (ants, bees, wasps). With 2 pairs membranous wings, hind wings smaller than forewings and interlocking with them by small hooks. Mouthparts primarily adapted for biting, but sometimes for lapping or sucking also. Abdomen usually constricted at base and its first segment fused with metathorax, forming a peduncle. Ovipositor always present and modified for stinging, piercing or sawing. Metamorphosis complete. Larvae generally without legs; head fairly well developed. Pupae generally in cocoon. Examples: *Nematus* (saw fly), *Apis* (honey bee), *Bombus* (bumble bee), *Vespa* (hornet) *Formica* (ant).

Order 8. Coleoptera (beetles). The largest insect (or any other) order, containing more than 220,000 species. Forewings modified as horny or leathery elytra meet in mid-line when folded and cover hind wings which are folded beneath them. Hind wings often reduced or absent. Biting mouthparts; ligula variably lobed. Prothorax large and mobile, mesothorax much reduced. Metamorphosis complete. Larvae generally with legs. Examples: *Goliathus* (goliath beetle, one of the largest insects), *Pyrophorus* (firefly), *Tenebrio* (flour beetle), *Tribolium*.

Order 9. Strepsiptera (stylopids). Endoparasites of other insects. Only male free-living, with mouthparts of degenerate biting type; conspicuous antennae with appearance of being bifurcated (though not in fact). Forewings reduced to small club-shaped structures; hind wings large and fan-shaped. Metathorax greatly developed. Female normally remains in host, enclosed in puparium which projects slightly from host's body; 3—5 pairs segmental genital openings; A few females leave hosts and have larviform structure and terminal gonopore. Example: *Stylops*.

8 Classification of Lophophorate Coelomates

Coelomates with a horseshoe-shaped structure (the lophophore) bearing ciliated tentacles, which serves as food collecting and respiratory organ. Body primitively divided into three regions: protosome, mesosome and metasome, the first being a pre-oral lobe (epistome), the second bearing the lophophore, and the third is the major part of the body containing the viscera. Ideally, each region of the body has a separate coelomic compartment: protocoel, mesocoel and metacoel, but the existence of the protocoel is in doubt, and in ectoprocts and brachiopods, the mesocoel and metacoel communicate with each other. Lophophorates are certainly related to one-another, but their relationship to protostomes and deuterostomes is uncertain because some of the critical embryological features of both are represented among the lophophorates.

Summary Classification of Lophophorate Coelomates

Phylum Phoronida
Phylum Ectoprocta
   Class 1 Phylactolaemata
   Class 2 Gymnolaemata
Phylum Brachiopoda
   Class 1 Inarticulata
   Class 2 Articulata

PHYLUM PHORONIDA

Tubiculous vermiform lophophorates, slender and solitary, living in secreted tubes. They possess metanephridia and closed blood vascular system, separate mesocoels and metacoels. Examples: *Phoronis, Phoronopsis* (only genera).

PHYLUM ECTOPROCTA

Very small sessile and colonial lophophorates, living permanently in a horny, calcareous, or gelatinous case secreted by themselves. They lack nephridia and a blood vascular system. Coelomic compartments of the mesosome and metasome confluent.

Class Phylactolaemata

Lophophore horseshoe-shaped, epistome present and the body wall with a muscle layer. Coelom of zooid in communication with that of neighbouring zooids. Case gelatinous, not calcified. Colonial animals living in freshwater. Examples: *Plumatella, Cristatella, Lophopus*.

Class Gymnolaemata

Lophophore circular, epistome and body-wall musculature missing. Direct coelomic communication between zooids in the colony. Nearly all marine.

Order 1. Ctenostomata. Case enclosing zooid membranous, not calcified, orifice terminal or sub-terminal. Examples: *Alcyonidium, Bowerbankia, Triticella, Paludicella* (the last in freshwater).

Order 2. Cheilostomata. Colonies branching, lamellate or encrusting. Zooids in more or less contiguous, box-like compartments, orifice sub-terminal and nearly always provided with a hinged lid. Case may be membranous or calcified. Examples: *Bugula, Membranipora, Flustra, Schizoporella*.

Order 3. Cyclostomata. Zooids tubular, more or less fused together and completely calcified, orifice terminal, circular without a hinged lid. Examples: *Crisia, Lichenopora, Tubulipora*.

PHYLUM BRACHIOPODA

Coelomate lophophorates with a bilaterally symmetrical bivalve shell, and generally attached to the substratum either directly or by a stalk. The dorsal and ventral valves of the shell are lined by the mantle lobe of the body wall and enclose the lophophore. Open circulatory system and 2 pairs of metanephridia. Extensive fossil record.

Class Inarticulata

The valves of the shell are held together by muscles only: lophophore without an internal supporting skeleton; possess an anus.

Order 1. Atremata. Stalk attached to the ventral valve, but both valves notched to allow its passage. Example: *Lingula* (lives in burrow, stalk not attached to substratum).

Order 2. Neotremata. Stalk emerges from notch or foramen in the ventral valve only. In *Crania* ventral shell of adult cemented directly to the substratum and stalk lacking. Examples: *Discinisca, Crania, Pelagodiscus*.

Class Articulata

Valves locked together by a tooth and socket at posterior margin; stalk emerges through a foramen in the ventral valve; lophophore generally with internal supporting skeleton; no anus. Ordinal classification of articulates is confused and at present only sub-orders (or super-families) are recognized.

Sub-order 1. Thecideoidea. With a simple lophophore indented in centre to form two lobes which fit into hollows in the dorsal valve. Examples: *Lacazella, Thecidellina*.

Sub-order 2 Rhynchonelloidea. Lophophore with two lobes, each elongated and coiled in helical spirals, without internal skeleton. With 2 pairs of nephridia. Example: *Hemithyris*.

Sub-order 3. Terebratuloidea. Lophophore with 2 simple, lateral lobes and a long, coiled central lobe; short supporting loop. Examples: *Terebratulina, Gryphus*.

Sub-order 4. Terebratelloidea. Lophophore with 2 simple, lateral lobes and a long, coiled central lobe; long supporting loop. Examples: *Terebratella, Megathyris*.

9 Classification of Deuterostome Coelomates

Metazoans with a true coelom (i.e. a secondary body cavity lying completely within the mesoderm) or derived from animals with a true coelom. The more primitive deuterostomes are oligomerous with the body divided into three regions: protosome, mesosome and metasome, each with its own coelomic compartment: protocoel, mesocoel and metacoel. The protocoel and mesocoel tend to be obliterated in the more advanced deuteostomes. Deuterostome characteristics are embryological and may be modified or obscured. In principle, cleavage of the eggs is radial, the blastopore becomes the anus, and the coelom is generally formed by enterocoely. The primitive larval form is supposed to be the dipleurula which has a circum-oral ciliary band as well as others encircling the body.

Summary Classification of Deuterostome Coelomates

Phylum Chaetognatha
Phylum Pogonophora
Phylum Hemichordata
    Class 1 Pterobranchia
    Class 2 Enteropneusta
    Class 3 Planktosphaeroidea
Phylum Echinodermata
  Sub-phylum Pelmatozoa
    Class 1 Crinoidea
  Sub-phylum Eleutherozoa
    Class 1 Holothuroidea
    Class 2 Echinoidea
      Sub-class 1 Perischoechinoidea
      Sub-class 2 Euechinoidea

    Class 3 Asteroidea
    Class 4 Ophiuroidea
Phylum Chordata
  Sub-phylum Urochordata
    Class 1 Ascidiacea
    Class 2 Thaliacea
    Class 3 Larvacea
  Sub-phylum Cephalochordata
  Sub-phylum Craniata (Vertebrata)
    Class 1 Agnatha
      Sub-class 1 Osteostraci
      Sub-class 2 Heterostraci
      Sub-class 3 Cyclostomata
    Class 2 Placodermi
    Class 3 Chondrichthyes
      Sub-class 1 Elasmobranchii
      Sub-class 2 Holocephali
    Class 4 Osteichthyes
      Sub-class 1 Acanthodii
      Sub-class 2 Actinopterygii
        Infraclass 1 Chondrostei
        Infraclass 2 Holostei
        Infraclass 3 Teleostei
      Sub-class 3 Crossopterygii
      Sub-class 4 Dipnoi
    Class 5 Amphibia
      Sub-class 1 Labyrinthodontia
      Sub-class 2 Lepospondyli
      Sub-class 3 Lissamphibia
    Class 6 Reptilia
      Sub-class 1 Anapsida
      Sub-class 2 Lepidosauria
      Sub-class 3 Archosauria
      Sub-class 4 Ichthyopterygia
      Sub-class 5 Sauropterygia
      Sub-class 6 Synapsida
    Class 7 Aves
      Sub-class 1 Archaeornithes
      Sub-class 2 Neornithes

Class 8 Mammalia
   Sub-class 1 Prototheria
   Sub-class 2 Allotheria
   Sub-class 3 Eotheria
   Sub-class 4 Theria
      Infraclass 1 Trituberculata
      Infraclass 2 Metatheria
      Infraclass 3 Eutheria

PHYLUM CHAETOGNATHA

Small, slender, pelagic animals. Body torpedo-shaped with 1—2 pairs lateral, horizontal fins and a rounded, horizontal tail fin. A set of grasping spines arise at each side of the head which can be enveloped dorsally and laterally by a hood composed of a fold of the body wall. No nephridia or circulatory system. Hermaphrodite. Relationship with other deuterostomes uncertain. The anus is formed at the site of the blastopore and the coelom is formed by enterocoely, but here the resemblance ends. The coelom is suppressed in the larva and it is not certain if the adult secondary body cavity corresponds with the embryonic septum betweeen the head and trunk. Secondary sub-division of the trunk 'coelom' behind the anus gives one, or sometimes several tail coeloms. A small phylum containing about 6 genera; classes and orders not recognized. Examples: *Sagitta, Spadella, Eukrohnia.*

PHYLUM POGONOPHORA

Solitary, extremely long and slender, tubicolous worms. Body divided into three regions each with a coelomic compartment formed by enterocoely. The protosome bears a tentacular apparatus. No digestive system, food particles filtered from the water by the tentacles evidently digested externally. Exclusively marine, most occurring in deep water.
Order 1. Athecanephria. Protosome generally fused with mesosome externally. Nephric ducts diverge to separate nephridiopores at sides of the body. Posterior part of the trunk without rows of adhesive papillae. Tentacles free. Examples: *Siboglinum, Oligobrachia.*
Order 2. Thecanephria. Protosome generally fused with mesosome externally. Nephric ducts converge to a medial nephridiopore. Posterior part of the trunk with regularly repeated transverse rows of adhesive

papillae. Tentacles often bound together by a membrane. Examples: *Lamellisabella, Spirobrachia, Polybrachia*.

PHYLUM HEMICHORDATA

Enterocoelous coelomates with body and coelom divided into three regions. The tentacular apparatus, if present, borne on the mesosome. Nervous system with a mid-dorsal centre in the mesosome and mid-dorsal and mid-ventral cords in the metasome. A pre-oral gut diverticulum, extending into the protosome. Pharyngeal gill slits generally present. At one time included in the Chordata as a sub-phylum, but now generally recognized as a separate, if possibly related, phylum.

Class Pterobranchia

Small tubicolous hemichordates, living in aggregations or colonies in secreted tubes. Gut 'U'-shaped. With tentaculated lophophore arising from the mesosome. Gill slits, if present, simple and consist only of a single pair.

Order 1. Rhabdopleurida. Form true colonies in which individuals are in organic continuity with one another. With two tentaculated arms. No gill slits. Example: *Rhabdopleura*.

Order 2. Cephalodiscida. Form aggregations but individuals not in organic continuity with one another. With 4—9 pairs of tentaculated arms, and with one pair of simple gill slits. Examples: *Cephalodiscus, Atubaria*.

Class Enteropneusta

Vermiform, solitary hemichordates with numerous pharyngeal gill slits. Gut straight. No lophophore, and coelomic compartment of meso-some vestigial. Examples: *Balanoglossus, Ptychodera, Protoglossus, Saccoglossus*.

Class Planktosphaeroidea

Spherical pelagic larva with arborescent ciliary bands over the surface. 'U'-shaped gut. Coelomic sacs occupy a small part of the gelatinous interior. Presumed to be the larva of an unknown hemichordate. Example: *Planktosphaera*.

PHYLUM ECHINODERMATA

Entercoelous coelomates constructed, as a rule, on a pentaradiate plan, and lacking a definite head or brain. Evidence that the coelom is divided into three regions, as in other deuterostomes, during embryological stages, but adult morphology highly modified. Calcareous skeleton of separate plates or ossicles lies immediately beneath the epidermis, often with external spines and protuberances. With a unique water vascular system of a coelomic nature, which in juvenile stages and generally in the adult also, communicates with the external medium through a pore or cluster of pores set in a calcareous plate (the madre-porite) and usually has small thin-walled projections (podia) passing through the body wall to the outside. Extensive fossil record and a number of classes and orders are now extinct. Only those with living representatives are listed below.

Division of the Echinodermata into two sub-phyla, as given here, is usual, but largely a matter of convenience and probably does not reflect evolutionary relationships. Alternative groupings of the classes have been proposed but none has yet won universal acceptance. The system used here is based on that given in Rothschild (1961) *A Classification of Living Animals* (Longmans, London).

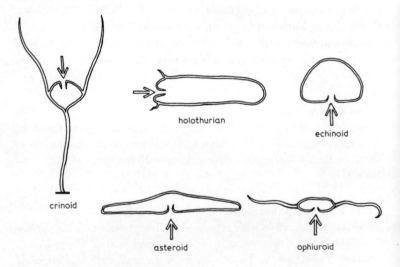

FIGURE 15   Body form and axis of symmetry in echinoderms.

SUB-PHYLUM PELMATOZOA

Echinoderms attached to the substratum as young forms, if not throughout life, by the aboral surface, either directly (in some extinct classes) or by a stalk (Fig. 15). Both mouth and anus are on the oral surface. Viscera enclosed in a calcareous test (theca). Podia are primarily food-catching and ciliated food grooves (ambulacra) usually extend distally from the mouth onto the projecting arms. Main nervous system aboral. Mostly extinct; only a single order of one of five classes still has living representatives.

Class Crinoidea

Theca differentiated into an aboral cup and a more or less disc-shaped oral roof. Pentamerously arranged arms are primitively simple but in most modern forms are branched; they are movable and attached to the theca at the junction between the aboral cup and the oral roof. Ambulacra extend along the whole length of the arms. Two sub-classes extinct and of the third, three of the four orders are extinct.

Order 1. Articulata. Aboral cup of theca flexible, oral roof leathery containing calcareous particles or small plates. Mouth and ambulacra exposed. Attached to substratum either as juveniles or throughout life by a stalk. Examples: *Antedon, Metacrinus, Comanthus, Neometra.*

SUB-PHYLUM ELEUTHEROZOA

Stemless, unattached echinoderms moving with the oral surface downwards, or lying on one side (Fig. 15). Ambulacra nervous and podia generally concerned with locomotion, not food gathering. Anus, when present, is on aboral surface. Main nervous system oral.

Class Holothuroidea

Elongated in the oral-aboral plane, with a secondary bilateral symmetry, and generally lying on one side. Body wall leathery with skeleton reduced to minute spicules or plates and sometimes lacking altogether. Mouth surrounded by a set of food-collecting tentacles connected with the water vascular system; podia locomotory or sometimes absent. Gonad a single or paired tuft of tubules. Often with a respiratory system consisting of a series of tubules arising just inside the anus (the respiratory trees) into which water can be drawn and expelled.

Order 1. Aspidochirota. With numerous podia. Oral tentacles with a short stem and finely branched at the tip. No oral retractor muscles. Respiratory trees present. Examples: *Holothuria, Stichopus.*

Order 2. Elasipoda. With numerous podia. Oral tentacles as in Aspidochirota. No oral retractor muscles or respiratory trees. Deepsea forms, often a few large papillae on the body surface. Examples: *Deima, Pelagothuria.*

Order 3. Dendrochirota. With numerous podia. Oral tentacles arborescent. Oral retractor muscles and respiratory trees present. Examples: *Cucumaria, Thyone.*

Order 4. Molpadonia. Without podia except for a few appearing as anal papillae. Oral tentacles finger-like. Posterior region generally tapers into a caudal portion. Respiratory trees present. Examples: *Molpadia, Caudina.*

Order 5. Apoda. Worm-like holothurians without podia and with a greatly reduced water vascular system. With oral tentacles but without respiratory trees. Example: *Synapta.*

Class Echinoidea

Skeletal calcareous plates closely fitting and generally fused together to form a rigid test. Body spheroidal, oval, heart-shaped or flattened to a disc. Covered with movable spines. The ambulacra extend from the mouth almost to the aboral pole, with the ambulacral nerve covered by skeletal plates (ambulacral plates). Podia almost exclusively locomotory, projecting through pores in the skeletal plates and flanking the ambulacra. Mouth and anus surrounded by flexible membranes (peristomial and periproctal membranes, respectively). Gonads pentamerous.

Systematics of the Echinoidea has undergone major revision in recent years. Although the older classification was acknowledged not to reflect probable relationships within the class, it still survives in a number of textbooks. The more recent classification is coming into general use and is given here.

Sub-class Perischoechinoidea

Symmetrical globular echinoids with up to 14 columns of plates in the interambulacra and 2—20 in the ambulacra; none of the plates compound. Without gills, sphaeridia or ophiocephalous pedicellariae (Fig. 16). (Sphaeridia are small, hard, transparent spheroids borne on

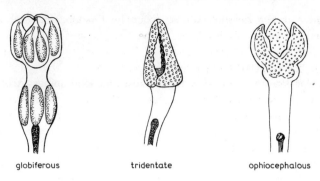

globiferous   tridentate   ophiocephalous

FIGURE 16   Echinoid pedicellariae.

the ambulacral areas.) Only one of four orders still has living representatives.

Order 1. Cidaroidea. Test rigid, globular, slightly flattened. Each interambulacral plate bearing one large primary spine which is surrounded at its base by a ring of small spines. Ambulacral and interambulacral plates continue across the peristome to the edge of the mouth. Examples: *Cidaris, Neocidaris.*

Sub-class Euechinoidea
Body shape variable. Ambulacra and interambulacra each with two columns of plates. Gills and sphaeridia present; ophiocephalous pedicellariae usually present.

SUPER-ORDER DIADEMATACEA
Globular sea-urchins with an Aristotle's lantern in which the teeth are grooved instead of having a keel (Fig. 17). The tubercles bearing the primary spines are perforated at their tip.

Order 1. Diadematoida. Rigid or flexible test. Primary spines hollow. Ten buccal plates on the peristomial membrane. Examples: *Diadema, Centrostephanus.*

Order 2. Echinothuroida. Flexible test with hollow primary spines. Simple ambulacral plates extend onto the peristomial membrane. Gills inconspicuous or lacking. Examples: *Asthenosoma, Phormosoma, Areosoma.*

Order 3. Pedinoida. Rigid test with solid primary spines. Ten buccal plates on the peristomial membrane. This group is sometimes included in the Echinothuroida. Example: *Caenopedina* (only living genus).

SUPER-ORDER ECHINACEA

Globular sea-urchins with a regular test and solid spines. Aristotle's lantern has keeled teeth. The pieces which carry the teeth (pyramids) are in two halves each with a projecting epiphysis at the aboral end. When the two ephiphyses of a pair are fused together, the lantern is termed camarodont (Fig. 17).

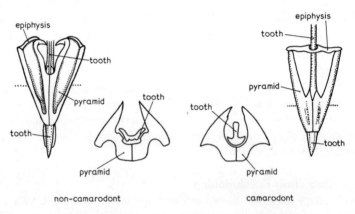

FIGURE 17   Tooth supports in jaw apparatus of regular echinoids: lateral and sectional views (after Hyman).

Order 1. Hemicidaroida. Lantern not camarodont. Tubercles of spines perforated. Most with a large plate in the periproctal region (suranal plate), which displaces the anus so that it lies eccentrically. Examples: *Salenia, Acrosalenia*.

Order 2. Phymosomatoida. Lantern not camarodont. Tubercles of spines not perforated. No suranal plate. Examples: *Glyptocidaris, Stomopneustes*.

Order 3. Arbacoida. Lantern not camarodont. Tubercles of spines not perforated. Periproct covered by 4—5 large plates. Example: *Arbacia*.

Order 4. Temnopleuroida. Lantern camarodont. Test usually sculptured with pits and depressions, but if not, the gill cuts are sharp and

deep. Examples: *Temnopleurus, Toxopneustes, Mespilia, Trigonocidaris, Lytechinus, Tripneustes, Sphaerechinus.*
Order 5. Echinoida. Lantern camarodont. Tubercles of spines not perforated. Test not sculptured and gill cuts shallow. Examples: *Echinus, Paracentrotus, Psammechinus, Echinometra, Strongylocentrotus.*

SUPER-ORDER GNATHOSTOMATA

Irregular urchins in which the mouth remains central on the oral surface but the anus has shifted out of the apical centre. The test is rigid and the lantern is retained. The aboral ambulacral areas may be modified and give the appearance of radiating petals ('petaloid') (Fig. 18). The oral ambulacral areas, if similarly modified, are known as phyllodes.

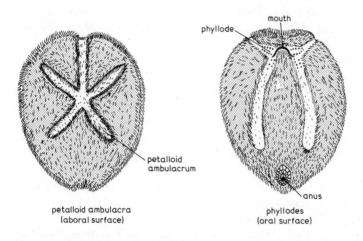

petalloid ambulacra
(aboral surface)

phyllodes
(oral surface)

FIGURE 18   An irregular echinoid: dorsal and ventral.

Order 1. Holectypoida. Mostly extinct. Test regular and ambulacra not petaloid, but periproct and anus displaced from the aboral apical system of plates and in living forms is on the oral surface. Examples: *Echinoneus, Micropetalon* (only living genera).
Order 2. Clypeasteroida. Sand dollars with test usually greatly flattened, with a round or oval outline. Aboral ambulacra petaloid, but no phyllodes, Example: *Clypeaster, Echinocyamus, Peronella, Mellita, Echinarachnius, Dendraster.*

SUPER-ORDER ATELOSTOMATA

Irregular urchins with a rigid test but without a lantern apparatus.
Order 1. Nucleolitoida. Periproct at posterior end of the test, 4—5 large genital plates in the apical system. Petals imperfectly developed or absent and no phyllodes. Example: *Neolampas*.

Order 2. Cassiduloida. Mostly extinct. Round or oval test. Displacement of periproct variable and may be central or slightly anterior. Phyllodes and imperfectly developed petals. Example: *Cassidulus*.

Order 3. Holasteroida. Elongated, oval or bottle-shaped urchins from deep water with a thin, fragile test. Petals and phyllodes not developed. Example: *Pourtalesia*.

Order 4. Spatangoida. Somewhat flattened heart urchins, rather elongated and bilaterally symmetrical. Four aboral ambulacra petaloid, the fifth, anterior, aboral ambulacrum not so. Phyllodes developed. Examples: *Echinocardium, Spatangus, Brissopsis, Moira, Meoma*.

Class Asteroidea

Flattened in the oral-aboral plane, mostly pentamerous, but sometimes six or more arms radiating symmetrically from the central disc. Mouth downwards. Ambulacra on the oral surface extending from the peristome to the tip of the arms, not covered by ossicles. Skeletal plates not fused and arms flexible. Gonads and digestive diverticula radially arranged and extending into the arms.

Order 1. Phanerozonia. Arms edged with conspicuous margin plates, in two rows, one on the oral, the other on the aboral surface, giving the arms some rigidity. Pedicellariae sessile. Examples: *Astropecten, Luidia, Goniaster, Porania*.

Order 2. Spinulosa. Not sharply distinguished from the Phanerozonia, but generally without marginal plates. Rarely have pedicellariae. Aboral surface with single spines or groups of spines. Podia with suckers. Examples: *Asterina, Henricia, Patiria, Solaster, Pteraster*.

Order 3. Forcipulata. Without conspicuous marginal plates, spines not in groups, pedicellariae all pedunculate. Podia with suckers. Examples: *Asterias, Pisaster, Pycnopodia*.

Class Ophiuroidea

Flattened, pentamerous echinoderms with long flexible arms, sometimes branched, sharply set off from the central disc. Ambulacral nerve completely covered by skeletal plates. No intestine or anus, madreporite on oral surface. Digestive system rarely extending into arms. Gonads pentamerous, discharging through invaginated sacs of the body wall (bursae), which are between the arms but sometimes are reduced or lacking.

Order 1. Ophiurae. Arms simple, move in a horizontal plane and relatively stiff (cannot twine around objects). Arm ossicles articulate by pit and projection joint. Arms and disc scaly. Examples: *Amphiura*, *Ophiothrix*, *Acrocnida*, *Ophiura*.

Order 2. Euryalae. Arms simple or branched, can move vertically and may be entwined around objects. Arm ossicles articulate by hour glass projections at right angles on adjacent faces. Scales on arms poorly developed and covered by a thick skin. Examples: *Gorgonocephalus*, *Asteronyx*.

PHYLUM CHORDATA

Deuterostome coelomates, but with little evidence of an oligomerous organization save for the transitory appearance of protocoel and metacoel in the embryonic stages of a few fishes and possibly cephalo-chordates. Characterized by the perforation of the pharyngeal walls by gill clefts, the possession of a hollow dorsal nerve cord, the central cavity of which communicates with the gut at its anterior end by a neurenteric canal, and by the possession of an axial skeleton in the form of a dorsal notochord lying immediately beneath the nerve cord. The anterior end of the central nervous system is dilated to form a brain. With a muscular post-anal tail which contains extensions of the nerve cord and notochord.

SUB-PHYLUM UROCHORDATA

Chordate features confined to the larval stages and are missing from the adult, except for pharyngeal gill clefts which are usually elaborate and hypertrophied, and open to a peripharyngeal cavity instead of directly to the exterior. The coelom develops late and is restricted to the pericardial cavity. Pharynx contains a ciliated mid-ventral gutter (the endostyle). Pharyngeal gill cleft system involved in food collection and respiration. No excretory organs.

Class Larvacea

Small pelagic urochordates in which the adult retains the larval characteristics of a muscular tail containing nerve cord and notochord. Generally supposed to be neotenous. A secreted test with sieve plates by which the animal traps food particles from a water current drawn through the test by undulations of the muscular tail. Two simple gill clefts. Three families, sometimes placed in a single order Copelata. Examples: *Oikopleura, Appendicularia, Fritillaria.*

Class Ascidiacea

Adults sessile and attached to the substratum. Solitary or colonial. Notochord and hollow, dorsal nerve cord in larva, but shed at metamorphosis and adult consists of an enlarged pharynx with complex gill clefts opening into a peripharyngeal cavity which leads to an exhalent chamber. Viscera in basal region. Secreted test closely invests the epidermis. Hermaphrodite.

Order 1. Enterogona. Solitary, or colonial and enclosed in a common test. Body often divided into two or three regions. Unpaired gonads in, or just in front of a loop of the intestine in the basal part of the animal. Oviduct and sperm duct run along the rectum and all three open in the exhalent (cloacal) chamber. Examples: *Clavellina, Polyclinum, Ascidia, Ciona, Phallusia.*

Order 2. Pleurogona. Solitary or colonial. Body never divided into separate regions. Gonads and intestine always situated on the side of the pharynx which has longitudinal folds. Examples: *Botryllus, Styela, Molgula.*

Class Thaliacea

Pelagic, barrel-shaped urochordates with the pharyngeal wall perforated by gill slits and forming a complete or partial diaphragm dividing the barrel into inhalent (buccal) and exhalent (cloacal) parts. Circular, or near-circular muscle bands in the body wall. Display alternation of generations with one developing from an egg (oozooid), the next vegetatively from a stolon (blastozooid); hence oozooids lack gonads.

Order 1. Pyrosomida. Colonial, in the form of a hollow cylinder with walls composed of the common test of the polyps which are arranged radially with respect to the colony. Polyp buccal cavity directed outwards, cloacal cavity opens into the central cavity of the colony. Diaphragm with gill slits complete as a rule. No larva. Oozooid insignificant, merely the first polyp of the colony. Example: *Pyrosoma*.

Order 2. Doliolida. With a 'tadpole' larva which metamorphoses, without settling, into the oozooid. Oozooid and blastozooid generations both prominent. Pharyngeal diaphragm never complete. Muscle bands form complete rings around the body. Example: *Doliolum*.

Order 3. Salpida. Development of oozooid direct, without interpolation of a larval phase. Oozooid and blastozooid generations both prominent. Pharyngeal diaphragm incomplete. Muscle bands form incomplete rings and often complicated. Examples: *Salpa*, *Thalia*.

SUB-PHYLUM CEPHALOCHORDATA

Metamerically segmented chordates with notochord and dorsal hollow nerve cord persisting in adult. Pharyngeal gill slits open into a peribranchial chamber (atrium) and not directly to the exterior. Atrium extends a considerable distance along body. Notochord extends to extreme anterior end of the body. Pharynx with a ventral endostyle and pharyngeal gill cleft system involved in food collection as well as respiration. Excretory organs protonephridia. Examples: *Branchiostoma* (= '*Amphioxus*'), *Asymmetron*.

SUB-PHYLUM CRANIATA (= VERTEBRATA)

Metamerically segmented chordates with a high degree of cephalization of the central nervous system to give a brain which is enclosed in a skeletal neuro-cranium that also houses the organs of special sense. Notochord and dorsal nerve cord also usually enclosed in skeletal tissue. Skeleton cartilage and/or bone, the latter originating phylogenetically in the dermatome. Pharyngeal gills respiratory if functional and not involved in food collection in the adult. Endostyle lacking in the adult condition (represented by the thyroid gland). Kidney of nephrons, functioning by ultra-filtration correlated with high blood pressure. Circulatory system closed, with a well-developed heart.

The basis of the taxonomic arrangement given here is that set out in A. S. Romer's. (1966) *Vertebrate Paleontology*: third edition. (Chicago University Press).

Class Agnatha

Branchial arches not modified to form jaws, notochord unrestricted, branchial skeleton in somatopleur (i.e. outside the coelom). Paired appendages pectoral only, or absent.

Sub-class Osteostraci

Exoskeleton of dermal bone, dorsal orbits and naso-hypophysial opening.

Order 1. Cephalaspida. Head shield consolidated, pectoral fins and heterocercal tail (Fig. 19) present in advanced forms, so-called electric organs on head shield. U. Silurian and Devonian. Examples: *Cephalaspis*, *Tremataspis*.

Order 2. Anaspida. Head shield of small scales, paired fins absent, tail hypocercal (Fig. 19). U. Silurian and Devonian. Example: *Birkenia*.

Sub-class Heterostraci

Exoskeleton of dermal bone, naso-hypophysial openings probably always paired, ventral, paired fins absent, tail hypocercal (Fig. 19).

Order 1. Pteraspida. Head shield of large plates, tending to fragment in later forms, single branchial opening on each side. M. Ordovician to Devonian. Example: *Pteraspis*.

Order 2. Coelolepida (Thelodonti). Whole body including head covered with small scales, gills probably with separate openings. M. Silurian to L. Devonian. Example: *Thelodus*.

Sub-class Cyclostomata

The living Agnatha. Bone secondarily absent, paired fins absent. The two groups may not be closely related.

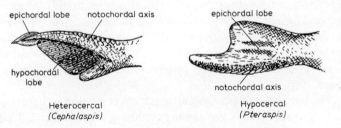

Heterocercal
(*Cephalaspis*)

Hypocercal
(*Pteraspis*)

FIGURE 19   The tail fin in Agnatha (after Stensiö and White).

Order 1. Petromyzontia (lampreys). Freshwater and anadromous forms. Eggs holoblastic with ammocoete larva. Naso-hypophysial opening dorsal; paired and pineal eyes functional. Roof of ventricles in brain non-nervous, ventricles large. Mouthparts suctorial, rasping. Examples: *Lampetra*, *Petromyzon*.

Order 2. Myxinoidea (hagfish). Marine, with large meroblastic eggs and abbreviated development. Naso-hypophysial opening terminal, eyes degenerate. Brain consolidated. Mouthparts excavating. Examples: *Myxine*, *Bdellostoma*.

The remaining vertebrates may be described as Gnathostomata, with the first gill arch incorporated into the braincase, the second (mandibular) forming jaws and the third (hyoid) probably always modified, originally for jaw suspension. Branchial skeleton in the splanchnopleur, pectoral and pelvic appendages normally present.

Class Placodermi

Heavy dermal armour of plates on head and thorax in primitive forms, individual dermal bones cannot be homologized with those of true bony fish, characteristic cranio-vertebral joint, thoracic spines, heterocercal tail. Jaw suspension possibly autostylic (Fig. 20) but not primarily so, no spiracle, gill arches in chamber in 'cheek' region.

Order 1. Arthrodira. Well-developed cranio-thoracic joint between skull and thoracic dermal armour, jaws of dermal bone with shearing edges, wide gape. Pectoral, pelvic and median fins. L. Devonian to L. Carboniferous. Example: *Coccosteus*.

Order 2. Antiarchi. Extensive thoracic armour, short head shield with dorso-medial eyes, nibbling mouth plates. Pectoral fins replaced by long articulated spines usually with an 'elbow' joint, pelvic fins sometimes present, 'soft anatomy' included spiral valve in gut and possibly paired lungs. L. to U. Devonian. Example: *Bothriolepis*.

A variable number of other placoderm orders, arguably related to arthrodires, are recognized and include sharklike, skatelike and chimaera-like forms.

Class Chondrichthyes

Bones secondarily absent except possible vestiges at base of dermal denticles. Spiral valve present, lungs or air bladder absent. Usually pelvic claspers as intromittent organs. Meroblastic eggs. The two

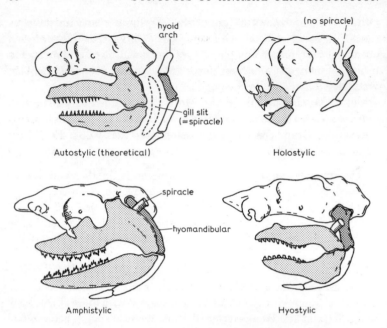

FIGURE 20   Suspension of the primary jaws in fish (diagrammatic).

major groups may not be closely related and are sometimes raised to
class status.

Sub-class Elasmobranchii

Amphistylic or hyostylic jaws (Fig. 20). Hyoid gill reduced to a spiracle,
remaining gills occupying separate pouches opening as slits to the
exterior. Tail fin heterocercal or reduced. Nostrils and mouth usually
ventral, thus rostrum present. Placoid 'scales' (dermal denticles)
(Fig. 21).

Order 1. Cladoselachii. Primitive broad based fin with basals parallel
to the axis. (Fig. 22). Claspers apparently absent. Jaws amphistylic.
Notochord apparently unconstricted. Devonian——Permian. Example:
Cladoselache.

Order 2. Selachii (sharks). Fins with constricted base. Claspers
present. Jaws amphistylic or hyostylic. Notochord normally constricted
by vertebrae. Devonian——Recent. Examples: *Heterodontus, Chlamy-
doselache, Scyliorhinus, Squalus.*

Order 3. Batoidea (skates and rays). Pectoral fins secondarily expanded as principal organs of locomotion. Gill openings (except spiracle) ventral in depressed body, tail and median fins reduced. Claspers and vertebrae present. Crushing dentition. Jurassic—Recent. Examples: *Rhinobatis*, *Pristis*, *Raja*, *Torpedo*.

Order 4. Xenacanthodii. Fins 'archipterygial' (Fig. 22), with pre- and postaxial radials in pectoral fins. Tail diphycercal. Claspers present. Jaws amphistylic. Long cephalic spine. U. Devonian—Triassic. Example: *Xenacanthus* (= *'Pleuracanthus'*).

Sub-class Holocephali

Jaws holostylic (Fig. 20), upper jaw fused to neurocranium, mouth and nostrils terminal, crushing dentition. Denticles much reduced, skin naked in living forms, tail whiplike. Gills covered by an operculum, spiracle closed. Numbers of ill-known fossil forms and the modern Chimaeras are included in this group. U. Devonian—Recent. Example: *Chimaera*.

Class Osteichthyes

Dermal bone covering head and forming basis of scales as in more primitive groups. Lungs or air bladder normally present and probably a primitive feature. Originally amphistylic or hyostylic, gills covered by an operculum. Pectoral girdle with a dermal component attached to skull. Tail primitively heterocercal.

It is now fashionable to raise the groups of Osteichthyes here termed sub-classes to class level. This partly solves the problem of the excessive number of intermediate taxonomic levels in the Actinopterygii but the difference between the primitive members of the sub-classes does not warrant such a wide separation.

Sub-class Acanthodii

An isolated group formerly placed with sharks, then placoderms. Fins preceded by spines, early forms have extra paired fins. Body and head covered with small scales. Amphistylic. U. Silurian——L. Permian. Examples: *Climatius*, *Acanthodes*.

Sub-class Actinopterygii

Hyostylic with sympletic bone in advanced forms. Neurocranium not kinetically divided. Incurrent and excurrent nostrils on side of snout, no choanae. Scales primitively ganoid (Fig. 21), suspension of jaws becomes mobile and migrates forward in evolution. Tail from heterocercal to homocercal (Fig. 23). Lung evolves to hydrostatic air bladder. Fin skeleton of radials not condensed into an axis of basals.

This vast and varied group of fish almost defies classification. The three infra-classes are admitted to be grades not clades and their names are now wildly inappropriate, although hallowed by general usage, in that they embody a theory of descent from cartilaginous fish.

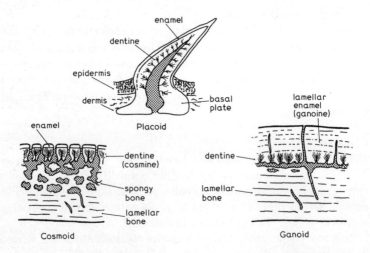

FIGURE 21   Scale types in fish in vertical section (diagrammatic—partly after Romer).

INFRA-CLASS CHONDROSTEI
Tail heterocercal or anomalous. Scales typically ganoid, fin skeleton with a broad series of radials at base. Spiracle retained, probably always spiral valve in gut. The most primitive actinopterygians and their aberrant living descendants.

Order 1. Palaeonisciformes. Ganoid scales, heterocercal tail without epichordal lobe. Large maxilla sutured to cheek and bordering mouth, jaws long. Primitively fusiform. Later forms ('subholostean') some freeing of maxilla. Devonian–L. Cretaceous. Examples: *Cheirolepis*, *Palaeoniscus*.

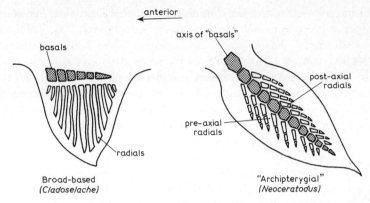

FIGURE 22   Pectoral fin skeleton in primitive fish (diagrammatic).

Order 2. Acipenseriformes. Degenerate living palaeoniscid derivatives. Cartilaginous skeleton, heterocercal tail. Skin naked or with a
few enlarged bony scales, sensory rostrum, weak suctorial mouths.
More ossified ancestors. Jurassic–Recent. Examples: *Acipenser*,
Polyodon.

Order 3. Polypterini (= Brachiopterygii). History unknown. Ventral lungs, fleshy lobe in paired fins. Tail externally symmetrical. Heavy
ganoid scales. Example: *Polypterus*.

INFRA-CLASS HOLOSTEI
A morphological grade between palaeoniscids and teleosts. Cosmine of
scales lost, ganoine reduced, tail hemi-homocercal. Maxilla shortened
and free of cheek posteriorly. Hyomandibular with symplectic bone in
some, spiracle lost. Reduction of dermal bones in cheek. Fin skeleton
reduced. May be divided into 5 or more orders. Triassic–Recent.

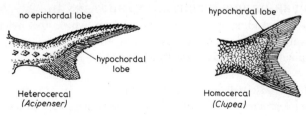

FIGURE 23   The tail fin in Actinopterygii.

Examples——extant: *Amia, Lepidosteus*; extinct: *Lepidotus, Pholidophorus*.

INFRA-CLASS TELEOSTEI

An enormous and complex group, possibly polyphyletic from the holostean level. Scales reduced to bone only, tail typically homocercal. Maxilla and premaxilla hinged at the front, the former toothless and underlain by the latter in advanced forms. Hyoid suspension highly mobile laterally. Supra-occipital in skull.

Classification of the 20—30,000 species of teleosts is highly complex and only super-orders are noted here. The first four super-orders are all primitive and possibly independently derived from the holostean level. Primitive characters include a maxilla still in the gape and bearing teeth, pelvic fins not advanced, a physostomatous swim-bladder (open to gut) and asymmetrical tail skeleton.

SUPER-ORDER 1. ELOPOMORPHA

Primitive or aberrant forms having a leaf-like leptocephalus larva. Jurassic—Recent. Examples: *Elops, Anguilla* (eel).

SUPER-ORDER 2. CLUPEOMORPHA

Primitive forms in which the swim-bladder extends forward to contact the inner ear. U. Jurassic or L. Cretaceous—Recent. Example: *Clupea* (herring).

SUPER-ORDER 3. OSTEOGLOSSOMORPHA

Primitive freshwater forms in which the primary bite is between the toothed parasphenoid and tongue. Cretaceous or Palaeocene—Recent. Examples: *Osteoglossum, Mormyrus*.

SUPER-ORDER 4. PROTACANTHOPTERYGII

Primitive, including generalized forms from which the remaining groups are probably derived. Cretaceous—Recent. Examples: *Salmo, Esox* (pike), *Gasterosteus* (sticklebacks).

SUPER-ORDER 5. OSTARIOPHYSI

The major group of freshwater teleosts, primitive forms in which the Weberian ossicles, formed from anterior vertebrae, connect swim-bladder and the inner ear. Palaeocene—Recent. Examples: *Cyprinus* (carp), *Phoxinus* (minnow), *Silurus* (catfish).

SUPER-ORDER 6. ATHERINOMORPHA
Primitive in position of pelvics etc. Advanced in toothless maxilla not in gape. Eocene—Recent. Examples: *Exocoetus* (flying fish), *Belone* (gar fish).

SUPER-ORDER 7. PARACANTHOPTERYGII
An advanced group, known to have an independent history from the next group and with primitive members not derivable from it. Like the next, some members develop spiny fin rays. Eocene—Recent. Examples: *Gadus* (cod), *Lophius* (angler fish).

SUPER-ORDER 8. ACANTHOPTERYGII
The major group of advanced teleosts, typically with spiny fins and ctenoid (spiny) scales. U. Cretaceous—Recent. Examples: *Perca* (perch), *Scomber* (mackerel), *Pleuronectes* (plaice), *Ostracion* (trunk-fish).

Sub-class Crossopterygii
Primarily hyostylic but without the forward migration of the suspensorium seen in Actinopterygii, neurocranium kinetically divided into anterior and posterior regions with notochord extending to the back of the anterior region. Tail heterocercal (with epicaudal lobe) to diphycercal (Fig. 24), fins derived from the archipterygial type (Fig. 22), two dorsals. Scales primitively cosmoid (Fig. 21).

Order 1. Rhipidistia. The only choanate fishes: includes tetrapod ancestors. Paired lungs presumed. Dermal roof, shoulder girdle and fins pretetrapod pattern. Teeth labyrinthine in section (Fig. 26). L. Devonian—L. Permian. Examples: *Holoptychius, Osteolepis, Eusthenopteron.*

Order 2. Actinistia (= Coelacanthina). No choanae. Maxilla and quadratojugal absent. Tail diphycercal. Air bladder ossified or reduced in later forms. M. Devonian—Recent. Examples: *Nesides, Undina* and the living *Latimeria.*

Sub-class Dipnoi
No choanae but excurrent nostril in mouth. Neurocranium consolidated. Premaxilla and maxilla absent, progressive reduction of other roofing

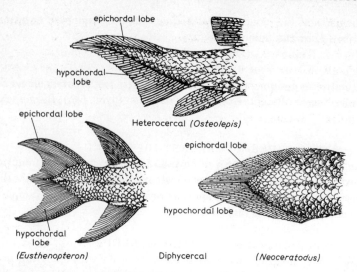

FIGURE 24   The tail fin in Crossopterygii and Dipnoi (partly after Jarvik).

bones. Holostylic, durophagous, crushing tooth-plates on pterygoids and prearticulars. Dermal roofing bones anomalous. Tail heterocercal to diphycercal (Fig. 24) continuous with median fins, paired fins archipterygial. Lung(s). Scales cosmoid originally. M. Devonian— Recent. Examples: *Dipterus*, *Sagenodus* and the living *Neoceratodus*, *Protopterus*, *Lepidosiren*.

Class Amphibia
Tetrapods in which internal gills have been lost but respiration by external gills and skin is common. Frequently a larval stage, and neoteny common. No amniote egg, lateralis system often present. Hyomandibular forms columella auris (stapes), secondarily autostylic. Choanae present.

Sub-class Labyrinthodontia
Centra of vertebrae originally with pleurocentra and intercentra. Skull with complete dermal roofing. Kinetism lost, stapes directed to a dorso-lateral tympanum. Teeth labyrinthine (Fig. 26). Cleithrum and inter-clavicle retained in pectoral girdle.
Order 1. Ichthyostegalia. Skull fishlike with pre- and subopercular bones, lateral line canals enclosed. External nares at margin of jaw.

Parasphenoid only below ethmoid neurocranium. Tail with fin rays.
U. Devonian. Example: *Ichthyostega.*

Order 2. Temnospondyli. Skull with postparietal contacting supra-temporal. Vomers broad: widely separated choanae. Vertebrae rhachi-tomous or, in late forms, stereospondylous (Fig. 25). Basicranial artic-ulation immobile in all but primitive forms, 4 fingers. Characterized by a series of evolutionary changes related to flattening of skull, etc. L. Carboniferous—U. Triassic. Examples: *Edops, Eryops, Tremato-saurus, Capitosaurus.*

Order 3. Anthracosauria (= Batrachosauria). Skull with tabular contacting parietal. Vomers narrow. Vertebrae embolomerous or gas-trocentrous (Fig. 25). Persistent basicranial articulation, 5 fingers. Remain primitive with respect to temnospondyl trends, but possibly include the ancestor of reptiles. L. Carboniferous—L. Permian. Examples: *'Palaeogyrinus'* (= *Palaeoherpeton*), *Archeria, Gephyros-tegus, Seymouria.*

Order 4. Plagiosauria. Skull roof as in temnospondyls. Dermal bones ornamented with pustules not a honeycomb pattern. Vertebrae anomalous, single elongate centra alternate with neural arches. Body

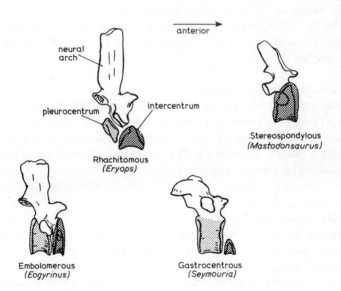

FIGURE 25   Diagnostic (trunk) vertebrate in the Labyrinthodontia.

armour of interlocking scutes. Body very wide and flat in advanced forms. U. Permian—U. Triassic. Examples: *Peltobatrachus*, *Gerrothorax*.

Sub-class Lepospondyli

Vertebrae with single spool-shaped centra. Dermal roofing bones often reduced or anomalous. Limbs sometimes reduced or even absent. Teeth simple.

Order 1. Aïstopoda. Limbs absent, pectoral girdle reduced. Roofing bones anomalous. Numerous vertebrae with reduced neural spines. L. Carboniferous—Permian. Examples: *Phlegothontia*, *Ophiderpeton*.

Order 2. Nectridea. Limbs developed, pectoral girdle with expanded cleithrum. Skull roof primitive, evolution of skull parallels that of Temnospondyli. Vertebrae with extra articulations, fan-shaped neural and haemal arches opposite on caudal intercentra. U. Carboniferous—L. Permian. Examples: *Batrachiderpeton*, *Diplocaulus*.

Order 3. Microsauria. Limbs developed but never more than 4 fingers, body form elongate or reptiliomorph. Haemal arches in tail alternate with centre. Contrast early reptiles: large supratemporal but no tabular in skull roof, occipital condyle concave, no transverse flange on pterygoid. Examples: *Microbrachis*, *Cardiocephalus*.

Order 4. Lysorophia. Possibly degenerate microsaurs. Dermal roof reduced. Limbs reduced. U. Carboniferous—L. Permian. Example: *Lysorophus*.

Sub-class Lissamphibia

The living Amphibia. Dermal roof of skull reduced, middle ear anomalous. Teeth pedicellate (Fig. 26). Larvae common. Lateralis system often retained. Anamniote. Scales reduced or absent, skin respiratory.

Order 1. Anura. Limbs and girdles developed for leaping, tail reduced. Larvae without true teeth and unlike adults. Frontals and parietals fused. Neoteny unknown. L. Triassic—Recent. Examples: *Ascaphus*, *Xenopus*, *Bufo*, *Hyla*, *Rana*.

Order 2. Urodela. Limbs normal or reduced, tail long. Sub-adult larvae with teeth. Frontals and parietals free. Neoteny common. Cretaceous—Recent. Examples: *Crytobranchus*, *Ambystoma*, *Salamandra*, *Triturus*, *Necturus*, *Siren*.

Order 3. Apoda. Limbless, tail reduced or absent. Meroblastic eggs, larval stage reduced or absent. Skull consolidated, eyes reduced.

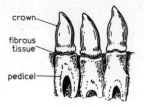

Labyrinthodont –T.S. at base of
crown *(Eogyrinus)*

Lissamphibia Pedicellate
(Gymnopis)

FIGURE 26 Teeth in Amphibia (after Atthey and Parsons & Williams).

Vestigial scales in some. Intromittant organ in male. No fossil record.
Example: *Hypogeophis*, *Gymnopis*.

Class Reptilia

Gills and larval stage never present. Lateral line system absent.
Amniote. Scales ectodermal, not homologous with those of fish. Meta-
nephric kidney. Temporal fenestration of the skull roof in more advanced forms. Single convex occipital condyle. Transverse flange on
pterygoid.

Sub-class Anapsida

Temporal openings in skull not normally developed (Fig. 27). Primitive
sprawling gait. Pectoral and pelvic girdles.

Order 1. Procolophonia. Primitive tetrapods. Large pineal opening. Wide otic notch. Skull often massive. Cleithrum retained in
pectoral girdle. Usually grouped with the next order as Cotylosauria. M.
Permian–U. Triassic. Examples: *Pareiasaurus*, *Procolophon*.

Order 2. Captorhinomorpha. Primitive reptiles. Pineal small, no
apparent otic notch. Skull lightly built: tabular and supratemporal
reduced. Cleithrum retained. Related to ancestry of probably all
advanced tetrapods. U. Carboniferous–L. Permian. Examples: *Captorhinus*, *Romeria*.

Order 3. Chelonia. Highly specialized. Vertebrae and ribs bearing
carapace and often fused to it, dermal pectoral girdle forming part of
plastron. Endochondral girdles thus within rib cage, cleithrum lost.
Skull massive, no pineal opening, frequently emarginated from behind.
U. Triassic–Recent. Examples: *Proganochelys*, *Testudo*, *Chelone*,
Chelus.

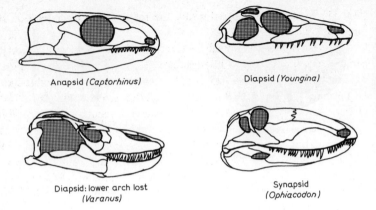

Anapsid *(Captorhinus)*

Diapsid *(Youngina)*

Diapsid: lower arch lost
(Varanus)

Synapsid
(Ophiacodon)

FIGURE 27   Reptile skulls to show temporal vacuities (after Fox & Bowman, Watson and Romer).

Sub-class Lepidosauria

Primitively diapsid: lower and then upper temporal arch may be lost (Fig. 27). Girdles and limbs unspecialized, often reduced or lost. Heart with partially divided ventricle, both auricles (as in Chelonia) open into one ventricle.

Order 1. Eosuchia. Lower temporal arch complete, quadrate fixed (monimostylic). Ancestral diapsids. The examples given show three possible stages in the evolution of the skull of Squamata and are sometimes separated as three orders. The first is not fully diapsid and may also be placed among the Anapsida. M. Permian—M. Triassic. Examples: *Milleretta, Youngina, Prolacerta.*

Order 2. Rhynchocephalia. Diapsid, lower arch normally complete, quadrate fixed. Postcranial skeleton primitive, intercentra in trunk vertebrae. Dentition acrodont (teeth fused to jaw). L. Triassic—Recent. Example: *Sphenodon* (the tuatara, the only surviving genus).

Order 3. Squamata. Lower arch always lost, upper sometimes; movable quadrate (streptostylic). Trunk intercentra rare. Limbs sometimes (lizards), or always reduced (snakes). L. Cretaceous (?Jurassic)—Recent. Examples: *Varanus, Iguana, Lacerta, Python, Natrix, Vipera.*

Sub-class Archosauria

Diapsid, arches complete. Trends associated with bipedalism in hind leg, pelvic girdle with elongate pubis and ischium, normally triradiate

(Fig. 28) reduction of dermal girdle and reduction of specialization of fore-limbs. Often large size.

Order 1. Thecodontia. Primitive forms. Skull with antorbital vacuity. Simple thecodont teeth (in sockets). Body armoured in midline. Acetabulum imperforate (Fig. 28). One sub-group crocodile-like. U. Permian–U. Triassic. Examples: *Euparkeria, Phytosaurus*.

Order 2. Crocodilia. Skull with secondary palate, no antorbital vacuity. Thecodont, simple teeth. Often armoured. Front limbs shorter than hind but elongate swimming body. Acetabulum perforate with pubis excluded (Fig. 28). 4 chambered heart, each auricle into its respective ventricle. U. Triassic–Recent. Examples: *Protosuchus, Geosaurus, Crocodylus, Alligator*.

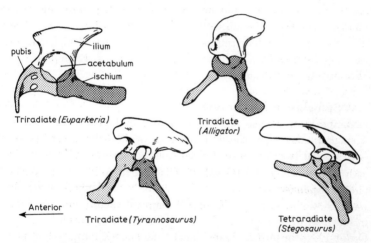

FIGURE 28   Pelvic girdles of archosaur reptiles in lateral view (after Ewer and Romer).

Order 3. Saurischia. Antorbital vacuity in skull. Teeth simple on premaxilla and maxilla and lower jaw. Dermal girdle reduced or absent. Triradiate pelvis, acetabulum perforate (Fig. 28). One of the 'dinosaur' orders. 2 sub-orders: bipedal carnivores and huge quadrupedal herbivores. U. Triassic–U. Cretaceous. Examples: (1) *Compsognathus, Megalosaurus, Tyrannosaurus*, (2) '*Brontosaurus*', *Diplodocus, Brachiosaurus*.

Order 4. Ornithischia. Antorbital vacuity small or absent. Teeth grinding, only in maxilla and dentary. Pre-dentary bone. Dermal girdle reduced or absent. Tetraradiate pelvis (Fig. 28),acetabulum usually open. The other 'dinosaurs'. All herbivorous, primitively partly bipedal, most quadrupedal. U. Triassic-U. Cretaceous. Examples: *Iguanodon, Stegosaurus, Scolosaurus, Triceratops*.

Order 5. Pterosauria. Flying reptiles. Skeleton pneumatic, patagium largely supported by 4th digit. No dermal girdle. No armour. Skull elongate, teeth lost in advanced forms, as is tail. L. Jurassic—U. Cretaceous. Examples: *Rhamphorhynchus, Pterodactylus, Pteranodon*.

Sub-class Ichthyopterygia

Order 1. Ichthyosauria. Marine reptiles, fusiform body with median dorsal and caudal fins and fin-like limbs with hyperphalangy (extra phalanges in digits). Skull with long rostrum and large orbits. Teeth numerous peg-like, labyrinthodont. A single pair of high temporal vacuities ('parapsid'). M. Triassic—U. Cretaceous. Examples: *Cymbospondylus, Ichthyosaurus*.

Sub-class Sauropterygia

Amphibious or marine reptiles, swimming by limbs rather than tail. A single pair of high placed vacuities as in the previous group, skull emarginated from below or derived from a diapsid condition, usually with reduction of quadratojugal. Teeth simple thecodont, neck elongate, girdles expanded ventrally. Various primitive and aberrant reptiles, some known to have a high placed temporal vacuity, are sometimes included with the Sauropterygia in a sub-class Euryapsida.

Order 1. Nothosauria. Transitional to an aquatic form, body slender, distal part of limbs (radius—ulna, tibia—fibula) shortened, some hyperphalangy. Skull with 'closed' palate—pterygoids meet in midline throughout. L.—U. Triassic. Example: *Nothosaurus*.

Order 2. Plesiosauria. Fully aquatic, barrel-shaped body, limbs converted to paddles with hyperphalangy. Skull with nares posterior, high placed; palate specialized but not to the same degree as nothosaurs. Two groups, one with short skull, long neck, the other vice-versa. M. Triassic—U. Cretaceous. Examples: (1) *Plesiosaurus, Elasmosaurus*, (2) *Pliosaurus*.

Order 3. Placodontia. Fully aquatic, body primitively nothosaur-like, later shorter and armoured with some having a turtle-like carapace.

Skull with posterior external nares, closed palate with enormous pebble-like teeth on palatine and along lateral jaw margins, anterior grasping 'incisors' or toothless beak. Sometimes separated as a sub-class. L.—U. Triassic. Examples: *Placodus, Placochelys*.

Sub-class Synapsida

'Mammal-like reptiles'. A single pair of lateral temporal vacuities (Fig. 27). Bridge the morphological gap between captorhinomorphs and early mammals. Primitively 2 coracoids.

Order 1. Pelycosauria. Primitive forms; temporal vacuity small, no false palate. Teeth usually homodont (uniform) but advanced forms bear canines. Primitive sprawling gait. M. Carboniferous—M. Permian. Examples: *Ophiacodon, Edaphosaurus, Dimetrodon*.

Order 2. Therapsida. Several parallel lines tending to mammalian condition. Temporal vacuities large, temporal dermal bones reduced. Quadrate and postdentary bones reduced. Some with heterodont dentition, and false palate. Postcranial skeleton approached mammalian condition and gait. M. Permian—U. Triassic. Examples: *Moschops, Dicynodon, Scymnognathus, Cynognathus, Bauria*.

Class Aves

Aerial descendants of thecodont archosaurs. Patagium and body covering of feathers. Skull with beak, teeth reduced or absent, streptostylic. Wing skeleton with reduced hand, large sternum with keel, clavicles retained. Pelvic girdle with long synsacrum, tetraradiate. Homoiothermic, heart of crocodile-type with loss of left systemic arch.

Sub-class Archaeornithes

Primitive, body elongate, tail retained, vertebrae amphicoelous (concave at both ends). Sternum without a prominent keel. Teeth retained, skull not fully streptostylic, endocranial cast reptilian. U. Jurassic. Example: *Archaeopteryx*.

Sub-class Neornithes

Body compact, tail reduced, vertebrae with saddle articulations. Sternum primitively with keel. Teeth normally absent, skull streptostylic, brain large with expanded cerebellum and optic lobes. The four super-orders are probably nearer ordinal rank.

SUPER-ORDER 1. ODONTOGNATHAE

Primitive aquatic birds, possibly retaining teeth. U. Cretaceous. Example: *Hesperornis*.

SUPER-ORDER 2. PALAEOGNATHAE

Possibly polyphyletic. Palate palaeognathous (Fig. 29). Mostly flightless forms without carina on the sternum. U. Cretaceous—Recent. Examples: *Struthio* (Ostrich), *Dromaius* (Emu), *Apteryx* (Kiwi), *Tinamus*.

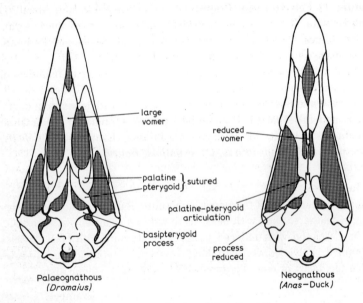

FIGURE 29   Bird skulls in palatal view (after Romer).

SUPER-ORDER 3. IMPENNAE

Penguins. Wings modified for swimming. Legs short, feet webbed. Adapted to cold climate. Eocene—Recent. Examples: *Spheniscus*, *Aptenodytes*.

SUPER-ORDER 4. NEOGNATHAE

The majority of birds, mostly flying. Sternum with carina, palate neognathous (Fig. 29). About 20 'orders'. Cretaceous-Recent. Example: *Corvus*, etc.

Class Mammalia

Culmination of the trends shown by synapsid reptiles. The reptile-mammal threshold is usually defined when a squamosal-dentary jaw articulation is achieved with incorporation of the reptilian articular (malleus), quadrate (incus) with the stapes as three ear ossicles, but there is an overlap even on this criterion between the most advanced mammal-like reptiles and the most primitive mammals. Skull with false palate and epipterygoid incorporated into braincase as alisphenoid, two occipital condyles etc. Dentition heterodont, diphyodont (single replacement). Dermal pectoral girdle reduced or lost, cleithrum and usually interclavicle gone, pelvic girdle with expanded ilium. Large brain. Usually homoiothermic—hair and sweat glands. Oviparous or usually viviparous, milk. Heart 4-chambered, single (left) systemic arch, diaphragm, etc.

Sub-class Prototheria

Order 1. Monotremata. Primitive and divergent from other mammals. Oviparous. Teeth reduced, pterygoids reptilian, no auditory bulla. Cervical ribs free, limbs and girdles primitive, interclavicle and 2 coracoids retained, humerus primitive. Epipubic bones. Cloaca retained. No pre-Pleistocene fossil records. Examples: *Ornithorhynchus, Tachyglossus*.

Sub-class Allotheria

Order 1. Multituberculata. Relationship to other mammals unknown. Molars with rows of bunodont cusps (Fig. 30), rodent-like incisors. Jugals lost, lacrimals reduced, no angular process. Basicranial region distinctive. Epipubic bones. U. Jurassic—L. Eocene. Examples: *Ctenacodon, Cimolomys*.

Sub-class Eotheria

Primitive mammals in which the molar teeth bear the three main cusps in a line or have a pattern derivable from this without rotation of cusps. Primitive members retain vestiges of reptilian post-dentary bones in a groove in the lower jaw. Braincase primitive where known, cochlea straight.

Order 1. Morganucodonta. Primitive 3-cusped molars, occlusion between upper and lower teeth imprecise. Jaw with an angular process far forward possibly related to a combined reptilian and mammalian jaw joint. U. Triassic—L. Jurassic. Example: '*Morganucodon*' (= *Eozostrodon*).

Order 2. Triconodonta. 3 principal cusps in straight line in molars (Fig. 30). Jaw without angle or groove for post-dentary bones. One sub-group may be true Theria. U. Jurassic—L. Cretaceous. Example: *Triconodon*.

Order 3. Docodonta. Jaw primitive as in morganucodonts with angular process. Anomolous transversely widened molars. U. Jurassic. Example:: *Docodon*.

Sub-class Theria

Molar teeth with pattern based on a triangle of cusps or derived from it. Spiral cochlea. The living theria are viviparous, have no interclavicle or anterior coracoid. Cervical ribs fused to vertebrae.

INFRA-CLASS TRITUBERCULATA

Mesozoic forerunners of living theria. Known principally from jaws and teeth.

Order 1. Symmetrodonta. Molar teeth with 3 or more cusps in a triangle and cingulum or shelf forming the base of the triangle (Fig. 30). No angular process to jaw. A Triassic form has postdentary bones. U Triassic—M. Cretaceous. Examples: *Kueneotherium, Tinodon*.

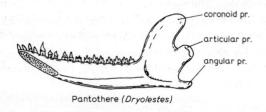

Pantothere *(Dryolestes)*

Multituberculate
(Ctenacodon)

Triconodont
(Triconodon)

Symmetrodont
(Tinodon)

Pantothere
(Amphitherium)

FIGURE 30   Lower jaw in mesial view and right lower molars in crown view of Mesozoic mammals (after Simpson and Crompton & Jenkins).

Order 2. Pantotheria. Ancestors of living Theria. Molars show the origin of the pattern in living forms, lower molars with talonid (a posterior shelf) (Fig. 30). True angular process (Fig. 30). M.–U. Jurassic. Examples: *Dryolestes*, *Amphitherium*.

INFRA-CLASS METATHERIA

Pouched mammals, young born underdeveloped and suckled in marsupium. Middle ear protected by alisphenoid. Epipubic bones. No corpus callosum in brain. Dental formula primitively

Incisors $\frac{5}{4}$ Canines $\frac{1}{1}$ Premolars $\frac{3}{3}$ Molars $\frac{4}{4}$ ($= 50$ for both sides of the skull).

U. Cretaceous–Recent. Examples: *Didelphis*, *Parameles*, *Macropus*.

INFRA-CLASS EUTHERIA (= PLACENTALIA)

Viviparous, with an allantoic placenta. Middle ear covered by petrosal. No epipubic bones. Corpus callosum between cerebral hemispheres.

Dental formula primitively $\dfrac{3. \ 1. \ 4. \ 3.}{3. \ 1. \ 4. \ 3.}$.

The problem of classifying placental mammals is analogous to that of classifying teleost fish. In the case of mammals, however, no grouping of the orders into larger taxa is satisfactory. Only orders with living members are diagnosed.

Order 1. Insectivora. Retain most features of primitive placentals. Commonly with primitive dental formula, molar teeth primitive often with reduction of the talonid. Skull with lacrimal bone extending onto face. Tympanic bulla if present formed at least in part from alisphenoid. Gait plantigrade (walking on the soles of the feet) (Fig. 31). U. Cretaceous–Recent. Examples: *Erinaceus* (hedgehog), *Sorex* (shrews), *Talpa* (moles), *Tenrec*.

Order 2. Dermoptera. Upper and lower incisors compressed, multicuspidate, the lower deeply pectinate. Dental formula $\dfrac{2. \ 1. \ 2. \ 3}{2. \ 1. \ 2. \ 3}$.

Orbit nearly surrounded by bone. Zygomatic arch well developed. Tympanic forms a bulla. Fore and hind limbs connected by a skin fold forming a parachute. Flying lemurs of Malaysia. Sumatra, Borneo and Phillipines. ?Paleocene–Recent. Example: *Cynocephalus* (= *Galeopithecus*) (the only living genus).

Order 3. Chiroptera. Fore limbs modified for flight, forearm with rudimentary ulna and long curved radius. Carpus with 6 bones sup-

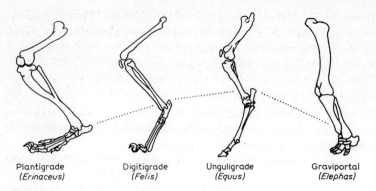

Plantigrade	Digitigrade	Unguligrade	Graviportal
(*Erinaceus*)	(*Felis*)	(*Equus*)	(*Elephas*)

FIGURE 31　Left hind limbs of mammals: relation of anatomy to gait.

porting a small pollex and 4 greatly elongated fingers, between which and the sides of the body and the hinder extremities, a patagium is supported. Hind limbs and other parts of the skeleton accordingly modified. Dental formula never exceeds $\frac{2.\,1.\,3.\,3.}{3.\,1.\,3.\,3.}$. Eocene—Recent. Bats. Examples: *Rhinolophus* (horse-shoe bat), *Vespertilio*, *Desmodus* (vampire bats), *Pteropus* (flying foxes).

Order 4. Primates. Limbs pentadactyl, arboreal, plantigrade, modified in different groups for climbing, leaping or brachiating, big toe and thumb opposible. Claws often replaced by nails. Progressive abbreviation of snout, enlargement of brain and elaboration of visual apparatus, often with stereoscopic vision. Olfactory sense reduced. Molars bunodont (Fig. 33) with simple 4 or 5 cusp pattern. Paleocene-Recent. Examples: ?*Tupaia* (tree shrew), *Galago* (bush baby), *Tarsius*, *Macaca*, *Gorilla*, *Homo*.

Order 5. Taeniodonta. Paleocene—Eocene.

Order 6. Tillodontia. Paleocene—Eocene.

Order 7. Creodonta. U. Cretaceous—L. Pliocene (archaic carnivores).

Order 8. Carnivora. In all but earliest forms, scaphoid, lunar and centrale of carpus fused (Fig. 32). Primitive dental formula probably $\frac{3.\,1.\,4.\,2.}{3.\,1.\,4.\,3.}$ as in dogs and bears. Canines strongly developed. Typically the last upper premolar (P^4) and the first lower molar (M_1) occlude together as shearing carnassials.

Typical land Carnivora show a wide radiation ranging from digitigrade (on the toes only) in specialized carnivores such as the cats, to plantigrade in herbivores such as the giant panda (Fig. 31). The aquatic seals etc. with simplified dentition and various degrees of modification of the limbs and teeth show the diagnostic carpal character and are included as a distinct sub-order. Paleocene—Recent. Examples: *Felis* (cats), *Canis* (dogs), *Ursus* (bears), *Procyon* (raccoon), *Meles* (badger), *Phoca* (seal), *Eumetopias* (sea-lion).

Order 9. Condylarthra. U. Cretaceous—U. Miocene (ancestral Ungulates).

Order 10. Amblypoda. Paleocene—Oligocene (archaic New World and Asian ungulates: possibly several unrelated groups).

Order 11. Proboscidea. In living elephants, skull high and pneumatic, head bearing trunk. Dental formula $\dfrac{1.\ 0.\ (3).\ 3.}{0.\ 0.\ (3).\ 3.}$. Upper incisors form tusks, position of lost lower incisors marked by 'chin'. Milk premolars and molars form dental battery in use in sequence from front to back, one or two each side in each jaw at a time, teeth hypsodont (Fig. 33) of a series of transverse plates with cement between. Skeleton massive, gait graviportal (weight bearing limbs) (Fig. 31), feet with cylindrical pad and nails. An exuberant radiation of fossil proboscideans is known with tusks in upper and/or lower jaw. Eocene—Recent. Examples: *Elephas* (Indian), *Loxodonta* (African).

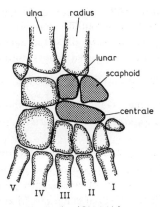

Primitive therian *(Didelphis)*

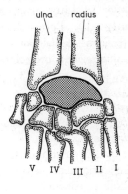

Carnivora *(Felis)*

FIGURE 32   Right carpus of a primitive therian mammal (after Romer) and a carnivore.

Order 12. Sirenia. Skull with heavy pachyostosis (thickening with obliteration of sutures), high curved premaxillary rostrum, nostrils far back, bones vestigial. Dentition aberrant and reduced. Incisors, canines reduced, dugongs have incisor tusks, in male reduced premolars and large bunodont molars. Manatees have above 20 cheek teeth with elephant-like replacement. Both herbivorous. Body naked, front limbs as paddles, reduced pelvic girdle, no apparent hind limbs. Eocene— Recent. Examples: *Halicore* (dugong), *Manatus* (manatee).

Order 13. Desmostylia. L. Miocene—L. Pliocene. (N. Pacific, sirenian-like skull, but with legs).

Order 14. Hydracoidea. Skull somewhat convergent to that of rodents or rabbits, a post-orbital bar usually present. Dental formula $\dfrac{1.\ 0.\ 4.\ 3.}{2.\ 0.\ 4.\ 3.}$. Upper incisors large and rootless, lower procumbent. Cheek teeth hypsodont with simple lophodont crowns (Fig. 33). Body rodent-like; plantigrade with digits over a pad and bearing nails as in elephants. L. Oligocene—Recent. Example: *Hyrax*.

Order 15. Embrithopoda. L. Oligocene. (only *Arsinoitherium* with a pair of massive nasal horns).

The following three orders constitute an enormous S. American radiation in some ways paralleling the perissodactyls.

Order 16. Notoungulata. Paleocene—Pleistocene.

Order 17. Astrapotheria. Eocene—Miocene.

Order 18. Litopterna. Paleocene—Pleistocene.

Order 19. Perissodactyla. Dentition of cropping incisors and hypsodont, lophodont cheek teeth (Fig. 33) with diastema (gap in canine region), premolars molarized, canines reduced. Dental formula $\dfrac{3.\ 1.\ 4.\ 3.}{3.\ 1.\ 4.\ 3.}$ (maximum for living forms). Axis of limbs through third digit, others variably reduced (mesaxonic). Gait unguligrade (hoofs) (Fig. 31). Eocene—Recent. Examples: *Equus* (horses), *Tapirus* (tapir), *Rhinoceros*.

Order 20. Artiodactyla. Premolars not molarized. Pattern of the limbs diagnostic, axis passes between third and fourth digit (paraxonic), variable reduction of outer digits, astragulus of ankle with pulley surface dorsally and ventrally. Gait digitigrade or unguligrade. Modern forms divisible into two major groups: (1) pigs and their allies, dentition reduced from $\dfrac{3.\ 1.\ 4.\ 3.}{3.\ 1.\ 4.\ 3.}$ with cropping incisors, defensive canines and bunodont molars. Usually four toes round a pad. (2) ruminants, upper

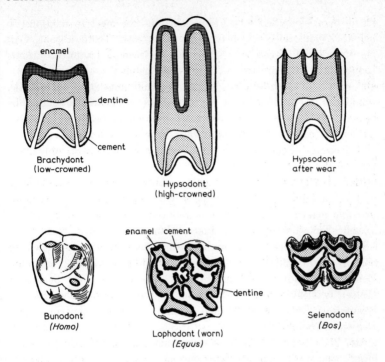

FIGURE 33   Above, mammalian molars, diagrammatic section (partly after Romer); below, upper molars in crown view.

incisors reduced to a single pair or lost, cropping lower incisors, a diastema with usually reduced canines and premolars, hypsodont, selenodont molars (Fig. 33). Horns or antlers often developed. Outer digits reduced or lost, third and fourth metacarpals and tarsals fused. Complex stomach. Eocene—Recent. Examples: (1) *Sus* (pig), *Hippopotamus*; (2) *Camelus*, *Giraffa*, *Cervus* (red deer), *Gazella*, *Bos* (cattle), *Ovis* (sheep).

Order 21. Edentata. Dentition reduced or absent, teeth without enamel. Variation in number of cervical vertebrae from 6—9. Extra (xenarthrous) articulartions between successive neural arches of posterior trunk vertebrae. Body often armoured. Gait plantigrade, arboreal, or derived from the latter. Paleocene—Recent. Examples: *Dasypus* (hairy armadillo), *Myrmicophaga* (great ant-eater), *Choloepus* (two-toed sloth).

Order 22. Pholidota. Body covered with imbricated scales. Skull in form of elongated cone, pterygoids do not meet in median line below. Zygomatic arch incomplete, jugal lost. Teeth absent. Gait plantigrade. Pleistocene (?Oligocene)—Recent. Example: *Manis* (pangolin or scaly ant-eater). Only known genus.

Order 23. Tubulidentata. Elongate skull, ant-eating, diagnostic, cylindrical teeth of dentine with vertical tubules, surrounded by cement, no enamel. 4—5 cheek teeth above and below in adult, no incisors or canines. Semi-plantigrade, digging forelimbs. L. Miocene (?L. Eocene)—Recent. Example: *Orycteropus* (aard-vark) only living genus.

Order 24. Cetacea. Whales: highly modified for swimming and unable to live on land. Forelimbs flippers with hyperphalangy, hind-limbs invisible externally, with skeleton reduced to a pair of bony rods or absent. Tail 'fin' of lateral flukes. Hair virtually lost, blubber for insulation. Skull in living forms highly modified with nostrils (blowhole) postero-dorsal, middle ear with vibratory bulla and reduced tympanum. Dentition of numerous peg-like teeth (toothed whales) or absent (baleen whales), filtering 'whalebone' in the latter. Eocene—Recent. Examples: *Balaenoptera* (blue whale), *Physeter* (sperm-whale), *Delphinus* (dolphin).

Order 25. Rodentia. The largest mammalian order. Single upper and lower incisors with persistently open roots and enamel only on anterior surface, thus self-sharpening, long diastema, cheek teeth hyposodont and lophodont. Dental formula (maximum) $\frac{1.\,0.\,2.\,3.}{1.\,0.\,1.\,3.}$ but premolars missing in advanced forms. Enlargement and specialization of the masseter muscle and its skull origin different in the principal sub-orders. Post-cranial skeleton generalized, plantigrade. Paleocene—Recent. Examples: *Sciurus* (squirrel), *Mus*, *Rattus*, *Hystrix* (porcupine), *Hydrochoerus* (capybara).

Order 26. Lagomorpha. Skull adaptations convergent to those of rodents, formerly classified in the same order or supra-ordinal group. Incisors open-rooted but with enamel all round, always a second small upper pair. Cheek teeth hypsodont, pattern simple by elongation of primitive cusps, formula $\frac{2.\,0.\,3.\,3.}{1.\,0.\,2.\,3.}$. Post-cranial skeleton typically modified for leaping with elongate tarsus. Tail reduced. U. Eocene (?Paleocene—Recent). Examples: *Oryctolagus* (rabbit), *Lepus* (hare), *Ochotona* (pika—short legs, short ears).

INDEX

Taxa below the rank of sub-order are not included. Where several page references are given, the diagnosis of the taxon will almost always be found on the last of these pages.

Acanthobdellae 51.
Acanthocephala 7, 30, 34.
Acanthocotyloidea 22, 25.
Acanthodii 81, 97.
Acari 55.
Acipenseriformes 99.
Acochlidiacea 42.
Acoela (Gastropoda) 43.
Acoela (Platyhelminthes) 22, 24.
Acoelomates 6, 22.
Acrothoracica 60.
Actiniaria 20.
Actinistia 101.
Actinopterygii 81, 97.
Actinulida 17.
Adenophora 30, 33.
Adepodonta 45.
Agnatha 81, 94.
Aïstopoda 104.
Alcyonacea 19.
Alcyonaria 19.
Alloeocoela 22, 24.
Allotheria 82, 111.
Amblypoda 115.
Amblypygi 54.
Ammonoidea 37, 46.
Amphibia 81, 102.
Amphidiscophora 10.
Amphilinoidea 23, 26.
Amphineura 36, 39.
Amphinomorpha 47.
Amphipoda 63.
Anamorpha 37, 66.

Anapsida 81, 105.
Anaspida 94.
Anaspidacea 62.
Anaspidea 42.
Anisomyaria 44.
Annelida 7, 37, 47.
Anomalodesmata 45.
Anopla 23, 29.
Anopleura 73.
Anostraca 57.
Antherinomorpha 101.
Anthomedusae 16.
Anthozoa 14, 18.
Anthracosauria 103.
Antiarchi 95.
Antipatharia 19.
Anura 104.
Aphasmidia 33.
Aplacophora 36, 40.
Aplysiomorpha 42.
Apoda (Amphibia) 104.
Apoda (Crustacea) 61.
Apoda (Echinodermata) 86.
Apterygota 37, 67.
Arachnida 37, 53.
Araneida 55.
Arbacoida 88.
Archaeogastropoda 40.
Archaeornithes 81, 109.
Archiacanthocephala 35.
Archiannelida 49.
Archosauria 81, 106.
Arthrodira 95.

Arthropoda 7, 37, 52.
Articulata (Brachiopoda) 77, 79.
Articulata (Echinodermata) 85.
Ascaroidea 33.
Aschelminthes 7, 30.
Ascidiacea 81, 92.
Ascorhynchomorpha 56.
Ascothoracica 61.
Aspidobothria 23, 28.
Aspidochirota 86.
Aspidogastrea 23, 28.
Asteroidea 81, 90.
Astrapotheria 116.
Atelostomata 90.
Athecanephria 82.
Athecata 16, 17.
Atremata 79.
Aves 81, 109.

Basommatophora 43.
Bathynellacea 62.
Batoidea 97.
Batrachosauria 103.
Bdelloidea 31.
Bdellomorpha 23, 29.
Beroida 21.
Bivalvia 36, 43.
Brachiopoda 7, 77, 78.
Brachiopterygii 99.
Branchiopoda 37, 57.
Branchiura 37, 60.
Bullomorpha 42.

Calanoida 59.
Calcarea 9, 10.
Calcaronea 11.
Calcinea 11.
Caligoida 60.
Camallanida 33.
Cambalida 66.
Capsaloidea 22, 25.
Captorhinomorpha 105.
Carnivora 114.
Carnosa 12.
Cassiduloida 90.
Cephalaspida 94.
Cephalaspidea 42.

Cephalobaenida 52.
Cephalocarida 37, 57.
Cephalochordata 81, 93.
Cephalodiscida 83.
Cephalopoda 36, 46.
Ceriantharia 19.
Ceriantipatharia 14, 18.
Cestida 21.
Cestoda 23, 26, 27.
Cestodaria 23, 26, 27.
Cetacea 118.
Chaetodermomorpha 40.
Chaetognatha 7, 80, 82.
Chaetonotoidea 31.
Cheilostomata 78.
Chelicerata 37, 53.
Chelonia 105.
Chilognatha 37, 65.
Chilopoda 37, 66.
Chimaericoloidea 22, 26.
Chiroptera 113.
Chitonida 40.
Chondrichthyes 81, 95.
Chondrophora 16, 17.
Chondrostei 81, 98.
Chordata 7, 81, 83, 91.
Chordeumida 65.
Choristida 12.
Chromadorida 33.
Cidaroidea 87.
Cirripedia 37, 60.
Cladocera 58.
Cladocopa 58.
Cladoselachii 96.
Clathrinida 11.
Clitellata 37, 49.
Clupeomorpha 100.
Clypeasteroida 89.
Cnidaria 6, 14.
Coelacanthina 101.
Coelenterates 6, 14.
Coelolepida 94.
Coenothecalia 20.
Coleoidea 37, 46.
Coleoptera 75.
Collembola 68.
Colobognatha 66.

Colossendeomorpha 55.
Conchostraca 58.
Condylarthra 115.
Copelata 92.
Copepoda 37, 59.
Corallimorpharia 20.
Coronatae 18.
Craniata 1, 81, 93.
Creodonta 114.
Crinoidea 80, 85.
Crocodilia 107.
Crossopterygii 81, 101.
Crustacea 37, 56.
Ctenophora 6, 14, 21.
Ctenostomata 78.
Cubomedusae 18.
Cumacea 62.
Cyclophyllidea 23, 28.
Cyclopoida 59.
Cyclostomata (Chordata) 71, 94.
Cyclostomata (Ectoprocta) 78.
Cydippida 21.

Decapoda (Cephalopoda) 46.
Decapoda (Crustacea) 64.
Demospongea 9, 12.
Dendrochirota 86.
Dermaptera 71.
Dermoptera 113.
Derocheilocarida 60.
Desmostylia 116.
Deuterostome coelomates 7, 80.
Diadematacea 87.
Diclybothrioidea 22, 26.
Diclidophoroidea 22, 26.
Dictyoptera 72.
Dicyemida 9, 13.
Digenea 23, 28, 29.
Dioctophymatida 34.
Diphyllidea 23, 27.
Diplopoda 37, 64.
Diplura 68.
Dipnoi 1, 81, 101.
Diptera 75.
Docodonta 112.
Doliolida 92.
Dorylaimida 33.

Drilomorpha 48.

Echinacea 88.
Echinodermata 7, 80, 84.
Echinoida 89.
Echinoidea 80, 86.
Echinothuroida 87.
Echiura 7, 36, 38.
Ectoprocta 7, 77, 78.
Edentata 117.
Elasipoda 86.
Elasmobranchii 81, 96.
Eleutherozoa 80, 85.
Elopomorpha 100.
Embioptera 71.
Embrithopoda 116.
Endoprocta 7, 30, 35.
Endopterygota 69, 73.
Enopla 23, 29.
Enoplida 34.
Enterogona 92.
Enteropneusta 80, 83.
Eoacanthocephala 35.
Eosuchia 106.
Eotheria 82, 111.
Ephemeroptera 69.
Epimorpha 37, 66.
Eucarida 63.
Eucestoda 26, 27.
Euechinoidea 80, 87.
Eumalacostraca 62.
Eumetazoa 6, 8.
Eunicemorpha 47.
Euphausiacea 64.
Euryalae 91.
Euryapsida 108.
Eurypterida 53.
Eutardigrada 51.
Eutheria 82, 113.
Exopterygota 69.

Flabelligerimorpha 49.
Forcipulata 90.

Gastropoda 36, 40.
Gatrotricha 7, 30, 31.
Gelatinosa 9, 10.

Geophilomorpha 66.
Glomerida 65.
Glomeridesmida 65.
Gnathiidea 63.
Gnathobdellae 51.
Gnathostomata (Craniata) 95.
Gnathostomata (Echinoidea) 89.
Gordioidea 34.
Gorgonacea 20.
Grylloblattodea 71.
Gymnolaemata 77, 78.
Gymnosomata 42.
Gyrocotyloidea 23, 26.
Gyrodactyloidea 22, 25.

Hadromerina 12.
Halichondrina 12.
Haplobothrioidea 23, 27.
Haplosclerina 13.
Harpacticoida 59.
Helminthomorpha 65.
Hemichordata 7, 80, 83.
Hemicidaroida 88.
Hemimetabola 69.
Hemiptera 73.
Heterocoela 11.
Heterocyemida 13.
Heterodonta 45.
Heteronemertini 23, 29.
Heteroptera 73.
Heterostraci 81, 94.
Heterotardigrada 51.
Hexactinellida 8, 9.
Hexasterophora 10.
Hirudinea 37, 50.
Holasteroida 90.
Holectypoida 89.
Holocephali 81, 97.
Holometabola 73.
Holostei 81, 99.
Holothuroidea 80, 85.
Homocoela 10.
Homoptera 73.
Hoplocarida 64.
Hoplonemertini 23, 29.
Hydracoidea 116.
Hydrocorallina 17.

Hydroida 16, 17.
Hydrozoa 14, 15.
Hymenoptera 75.

Ichthyopterygia 81, 108.
Ichthyostegalia 102.
Impennae 110.
Inarticulata 77, 78.
Insecta 37, 67.
Insectivora 113.
Isopoda 63.
Isoptera 72.

Julida 66.
Juliformia 66.

Keratosa 9, 13.
Kinorhyncha 7, 30, 31.

Labyrinthodontia 81, 102.
Lagomorpha 118.
Lamellibranchia 36, 44.
Larvacea 81, 92.
Lecanicephaloidea 23, 27.
Lepidopleurida 39.
Lepidoptera 74.
Lepidosauria 81, 106.
Lepospondyli 81, 104.
Leptomedusae 17.
Leptostraca 61.
Lernaeopodoida 60.
Leucettida 11.
Leucosoleniidae 11.
Limacomorpha 65.
Limnomedusae 16, 17.
Linguatulida 51.
Lobata 21.
Lipostraca 58.
Lissamphibia 81, 104.
Lithobiomorpha 66.
Litopterna 116.
Lophophorate coelomates 7, 77.
Lysorophia 104.

Macrodasyoidea 31.
Madreporaria 20.
Malacostraca 37, 61.

Mallophaga 73.
Mammalia 82, 111.
Mandibulata 37, 56.
Mecoptera 74.
Mermithoidea 34.
Merostomata 37, 53.
Mesogastropoda 41.
Mesozoa 1, 6, 8, 9, 13.
Metatheria 82, 113.
Metazoa 1, 6, 8, 14.
Microsauria 104.
Milleporina 16, 17.
Mollusca 7, 36, 38.
Molpadonia 86.
Monaxonida 9, 12.
Monhysterida 33.
Monogenea 22, 25.
Monogononta 31.
Monopisthocotylea 22, 25.
Monoplacophora 36, 39.
Monotremata 111.
Monstrilloida 59.
Morganucodonta 111.
Multituberculata 111.
Myodocopa 58.
Mysidacea 62.
Mystacocarida 37, 60.
Myxinoidea 95.
Myxospongida 12.
Myzostomaria 37, 49.

Narcomedusae 17.
Natantia 64.
Nautiloidea 37, 46.
Nebaliacea 61.
Nectonematoidea 34.
Nectridea 104.
Nematoda 7, 30, 32.
Nematomorpha 7, 30, 34.
Nemertini 23, 29.
Neogastropoda 41.
Neognathae 110.
Neomeniomorpha 40.
Neoptera 69, 70.
Neornithes 81, 109.
Neotremata 79.
Neritacea 41.

Neuroptera 74.
Nippotaeniidea 23, 28.
Notaspidea 42.
Nothosauria 108.
Notodelphoida 59.
Notostraca 58.
Notoungulata 116.
Nucleolitoida 90.
Nuda (Ctenophora) 14, 21.
Nuda (Porifera) 8, 9.
Nudibranchia 43.
Nymphonomorpha 55.

Octocorallia 14, 19.
Octopoda 46.
Odonta 69.
Odontognathae 110.
Oligochaeta 37, 50.
Oligoneoptera 69, 74.
Oniscomorpha 65.
Onychophora 7, 37, 52.
Ophiurae 91.
Ophiuroidea 81, 90.
Opiliones 55.
Opisthobranchia 36, 41.
Opisthopora 50.
Ornithischia 107.
Orthonectida 9, 13.
Orthoptera 71.
Ostariophysi 100.
Osteichthyes 1, 81, 97.
Osteoglossomorpha 100.
Osteostraci 81, 94.
Ostracoda 37, 58.
Osteichthyes 1.
Oweniimorpha 48.
Oxyuroidea 33.

Palaeacanthocephala 35.
Palaeognathae 110.
Palaeonemertini 23, 29.
Palaeonisciformes 98.
Palaeoptera 69.
Palpigradi 54.
Pancarida 62.
Pantotheria 113.
Paracanthopterygii 101.

Paraneoptera 69, 72.
Parazoa 1, 6, 8.
Pauropoda 37, 67.
Pedinoida 87.
Pelycosauria 109.
Pelmatozoa 80, 85.
Pennatulacea 20.
Pentastomida 7, 37, 51.
Pentazonia 65.
Peracarida 62.
Perischoechinoidea 80, 86.
Perissodactyla 116.
Petromyzontia 95.
Phanerozonia 90.
Pharetronida 11.
Pharyngobdellae 51.
Phasmida 71.
Phasmidia 32.
Pholidota 117.
Phoronida 7, 77.
Phthiraptera 73.
Phylactolaemata 77, 78.
Phyllocarida 61.
Phyllodocemorpha 48.
Phymosomatoida 88.
Placentalia 113.
Placodermi 81, 95.
Placodontia 108.
Plagiosauria 103.
Planktosphaeroidea 80, 83.
Platycopa 59.
Platyctenia 21.
Platyhelminthes 6, 22, 23.
Plecoptera 70.
Plesiopora-Plesiothecata 50.
Plesiopora-Prosothecata 50.
Plesiosauria 108.
Pleurobranchomorpha 42.
Pleurogona 92.
Podocopa 58.
Poecilosclerina 12.
Pogonophora 7, 80, 82.
Polychaeta 37, 47.
Polycladida 22, 24.
Polydesmida 65.
Polyneoptera 69, 70.
Polyopisthocotylea 22, 25.

Polyplacophora 36, 39.
Polypterini 99.
Polystomatoidea 22, 26.
Polyxenida 65.
Porifera 6, 8, 9.
Porocephalida 52.
Priapulida 7, 30, 35.
Primates 114.
Proacanthopterygii 100.
Proboscidea 115.
Procolophonia 105.
Prosobranchia 36, 40.
Prosopora 50.
Protista 1, 8.
Proteocephaloidea 23, 28.
Protobranchia 36, 44.
Protogyrodactyloidea 22, 25.
Protostome coelomates 7, 36.
Prototheria 82, 111.
Protozoa 1.
Protura 68.
Psammodrilomorpha 49.
Pselaphognatha 37, 64.
Pseudocoelomates 7, 30.
Pseudophyllidea 23, 27.
Pseudoscorpiones 54.
Psocoptera 72.
Pteraspida 94.
Pterobranchia 80, 83.
Pterosauria 108.
Pterygota 37, 68.
Pulmonata 36, 43.
Pycnogonida 37, 55.
Pycnogonomorpha 56.
Pyrosomida 92.

Reptantia 64.
Reptilia 81, 105.
Rhabditida 32.
Rhabditoidea 33.
Rhabdocoela 22, 24, 25.
Rhabdopleurida 83.
Rhipidistia 101.
Rhizocephala 61.
Rhizostomeae 18.
Rhynchobdellae 51.
Rhynchocephalia 106.

Rhynchocoela 6, 23, 29.
Rhynchonelloidea 79.
Ricinulei 55.
Rodentia 118.
Rotifera 7, 30.

Sacoglossa 42.
Sarcopterygii 1.
Saurischia 107.
Sauropterygia 81, 108.
Scaphopoda 36, 43.
Schizodonta 45.
Scleractinia 20.
Scolopendromorpha 66.
Scorpionones 54.
Scutigeromorpha 67.
Scyphozoa 14, 18.
Secernentia 30, 32
Seisonidea 31.
Selachii 96.
Semaeostomeae 18.
Septibranchia 45.
Serpulimorpha 49.
Siphonaptera 75.
Siphonophora 17.
Siphunculata 73.
Sipuncula 7, 36, 38.
Sirenia 116.
Solpugida 54.
Spatangoida 90.
Spelaeogriphacea 63.
Spinulosa 90.
Spiomorpha 48.
Spirobolida 66.
Spirostreptida 66.
Spirostreptomorpha 66.
Spirurida 33.
Squamata 106.
Stauromedusae 18.
Sternaspimorpha 48.
Stolonifera 20.
Stomapoda 64.
Strepsiptera 75.
Strongyloidea 33.
Stylasterina 16, 17.
Stylommatophora 43.
Sycettida 12.

Symmetrodonta 112.
Symphyla 37, 67.
Synapsida 81, 109.
Syncarida 62.

Taeniidea 23, 28.
Taeniodonta 114.
Tanaidacea 63.
Tardigrada 7, 37, 51.
Taxodonta 44.
Teleostei 81, 100.
Telestacea 20.
Temnocephalidea 22, 25.
Temnocephaloidea 22, 24.
Temnopleuroida 88.
Temnospondyli 103.
Tentaculata 14, 21.
Terebellomorpha 49.
Terebratelloidea 79.
Terebratuloidea 79.
Tetrabothrioidea 23, 27.
Tetractinellida 9, 12.
Tetraphyllidea 23, 27.
Tetrarhynchoidea 23, 27.
Thaliacea 81, 92.
Thecanephria 82.
Thecata 16, 17.
Thecideoidea 79.
Thecodontia 107.
Thecosomata 42.
Therapsida 109.
Theria 82, 112.
Thermosbaenacea 62.
Thoracica 60.
Thysanoptera 73.
Thysanura 68.
Tillodontia 114.
Trachylina 17.
Trachymedusae 17.
Trematoda 23, 28.
Trichoptera 74.
Trichosyringida 34.
Trichurata 34.
Tricladida 22, 24.
Triconodonta 112.
Trilobitomorpha 37, 52.
Trituberculata 82, 112.

Tryblidioidea 39.
Trypanorhyncha 23, 27.
Tubulidentata 118.
Turbellaria 22, 23, 25.
Tylenchida 32.
Tylenchoidea 33.

Udonelloidea 22, 25.
Urochordata 81, 91.
Urodela 104.
Uropygi 54.

Vampyromorpha 46.
Vertebrata 1.

Xenacanthodii 97.
Xiphosura 53.

Zeugloptera 74.
Zoantharia 14, 20.
Zoanthidea 20.
Zoanthiniaria 20.
Zoraptera 72.